Environmental Planning
and Sustainability

Environmental Planning and Sustainability

Edited by

Susan Buckingham-Hatfield

Brunel University College, London

and

Bob Evans

South Bank University, London

JOHN WILEY & SONS

Chichester · New York · Toronto · Brisbane · Singapore

Copyright © 1996 by John Wiley & Sons Ltd,
Baffins Lane, Chichester,
West Sussex PO19 1UD, England

National   01243 779777
International   (+44) 1243 779777

Other Wiley Editorial Offices

John Wiley & Sons, Inc., 605 Third Avenue,
New York, NY 10158-0012, USA

Jacaranda Wiley Ltd, 33 Park Road, Milton,
Queensland 4064, Australia

John Wiley & Sons (Canada) Ltd, 22 Worcester Road,
Rexdale, Ontario M9W 1L1, Canada

John Wiley & Sons (Asia) Pte Ltd, 2 Clementi Loop #02-01,
Jin Xing Distripark, Singapore 0512

Library of Congress Cataloging-in-Publication Data

Environmental planning and sustainability/edited by Susan Buckingham
   –Hatfield and Bob Evans.
      p. cm.
   Includes bibliographical references and index.
   ISBN 0–471–96480–8 (pbk.)
   1. Sustainable development—Congresses.   2. Environmental policy–
–Congresses.   I. Buckingham-Hatfield, Susan.   II. Evans, Bob, 1947–
HC79.E5E578456 1996
363.7—dc20   95–51037
                                                        CIP

British Library Cataloguing in Publication Data

A catalogue record for this book is available from the British Library

ISBN 0–471–96480–8

Typeset in 11/13pt Palatino by Dorwyn Ltd, Rowlands Castle, Hants
Printed and bound in Great Britain by Biddles Ltd, Guildford and King's Lynn

This book is printed on acid-free paper responsibly manufactured from sustainable
forestation, for which at least two trees are planted for each one used for paper production.

Contents

About the Contributors

Julian Agyeman is an environmental consultant and writer. A teacher of geography and environmental studies in the early 1980s, he became an environmental education adviser at Notting Dale Urban Studies Centre; Lambeth Council and Islington Council. In 1988 he was founder, and chair until 1994, of the Black Environment Network. He is co-editor, with Bob Evans, of *Local Environmental Policies and Strategies* (Longman 1994) and co-editor of the journal *Local Environment* (Carfax Publishing), and is presently completing a PhD (An Alternative Approach to Urban Nature in Environmental Education at Key Stage 2) at the Institute of Education, University of London.

Duncan Bayliss has a geographical and planning background. He has worked in local government in land use planning and has also undertaken a variety of research projects on planning-related topics. He now lectures on environmental management in the School of Planning at the University of the West of England, Bristol. His recent research has involved projects investigating environmental monitoring methodologies across the EU, access to environmental information, and state-of-the-environment reporting, and is currently developing improved access to environmental information for third parties. Duncan is also researching material for a comparative critique of the concept of sustainability from different presuppositional perspectives, especially a Christian perspective.

Susan Buckingham-Hatfield lectures in geography and environmental issues at Brunel University, West London. Her recent research has been into women's and men's attitudes and behaviour towards environmental issues and how participation in environmental policy-making is gendered. As a consultant to CSVEducation, she is involved in developing links between higher education and the community and voluntary sector which aim to increase the resources available to the community, with particular reference to environmental education. Susan is currently Secretary of the Planning and Environment Study Group of the RGS with IBG.

Bob Evans is Head of Geography and Environment at South Bank University, London. He is author of *Experts and Environmental Planning* (Avebury 1995) and co-author (with Julian Agyeman) of *Local Environmental Policies and Strategies* (Longman 1994). He is co-editor (also with Julian Agyeman) of the journal *Local Environment*, and has worked as a town planner in the public, private and voluntary sectors.

Tim Marshall is a lecturer in the School of Planning, Oxford Brookes University. He was trained as a planner and worked for several years in planning practice in Birmingham and London. His doctorate at Oxford Brookes University was on local planning and urban renewal in England. Present major interests include the environmental dimensions of planning and of economic change, both in Britain and in other European countries, especially at a regional level. After research in Spain in 1990–91, he has continued to work on processes of urban and environmental change in Barcelona, Catalonia and Spain as a whole.

Judith Matthews is Principal Lecturer in the Department of Geographical Sciences, University of Plymouth. She is a social geographer with research interests in the social processes resulting from people's relationship with their immediate physical surroundings. This leads to a focus on understanding how people react to changes in the environment at all scales from the neighbourhood to the global setting. A committee member, and recently chair of the Planning and Environment Study Group of the RGS with IBG, she has been involved in promoting the international comparative dimension of the group's activities, particularly in relation to environmental and urban planning.

Duncan McLaren is currently Research Coordinator of the Sustainable Development Unit of Friends of the Earth Trust Ltd, conducting and managing research in sustainable development issues. His current research interests include sustainability principles, targets and indicators, policy instruments and policy integration. He is a member of the steering committee of Friends of the Earth International's 'Sustainable Europe' project, in which 30 groups across Europe are developing sustainability plans based on the concept of environmental capacity.

Mary Mellor is Professor of Sociology at Northumbria University, Newcastle-upon-Tyne. Author of *Breaking the Boundaries* (Virago 1992) and many articles and papers on eco-feminism, she is currently writing a book on *Feminism and Ecology* to be published by Polity Press. She has also written extensively on the co-operative movement. These interests are brought together in her political commitment to the development of alternative socio-economic systems based on feminist, green and egalitarian values.

George Myerson is Reader in English at King's College London and writes on rhetoric and argument; his most recent book is *Rhetoric, Reason and Society*, published by Sage.

Yvonne Rydin is Senior Lecturer in Human Geography at the London School of Economics and Coordinator of the RICS Environmental Research Programme; she writes widely on the planning system, most recently in relation to sustainability, and is the author of *The British Planning System*, published by Macmillan. Since 1989, Yvonne Rydin and George Myerson have developed an interdisciplinary project on the rhetoric and discourse of the planning process and the wider environmental agenda. The work appears in *Society and Space, Planning Practice and Research* and a forthcoming book to be published in 1996 by UCL Press.

Ben Tuxworth has 13 years' experience of environmental work and currently runs the Environment Resource and Information Centre (ERIC) at the University of Westminster. He has worked as a consultant in Scotland, England and overseas, advising local authorities, government agencies, voluntary bodies and other groups on environmental programmes, particularly environmental education, interpretation and training. His work at ERIC focuses on the work of environment co-ordinators in local authorities and Local Agenda 21 UK. ERIC provides resources, training and good practice information to environment co-ordinators in local government, and publishes the monthly environmental newsletter, *EG*.

Gordon Walker is a principal lecturer in Geography and head of Environment and Resource Management at Staffordshire University. He has written and researched in a number of areas of environmental planning and policy, including land use planning and industrial hazard accidents, public perceptions of technological risks, environmental monitoring and the implementation of environmental legislation. Undertaking research for government agencies in the UK and European Union has enabled him to make direct inputs to the development of recent environmental policy and practice. Gordon is currently Chair of the Planning and Environment Study Group of the RGS with IBG.

Preface

The idea for this book emerged from one of the first critical conference sessions to be organised on the issue of how sustainability might be achieved through environmental planning. This book's editors, under the auspices of The Planning and Environment Study Group of the Institute of British Geographers with the Royal Geographical Society, convened a session at the Institute's conference at Royal Holloway and Bedford College which brought together academics, planning practitioners, non-governmental organisations and politicians to debate the meaning of sustainability and to propose ways of achieving this through the policy processes of environmental planning.

Whilst we did not canvass a particular approach, we hoped, through our choice of speakers, to emphasise the links between sustainability and democracy whilst addressing the conflicts this dual project is likely to produce. And they did not disappoint us. Although not all speakers have contributed to this book (we have tried to avoid the more overtly polemical positions), and others have been invited to do so where gaps in coverage were perceived – the themes which dominated the session are strongly reinforced through this volume.

The following chapters, whilst highlighting the semantic and practical ambiguities of sustainability, argue that this should not detain us in implementing a form of environmental planning that will enable us to work towards a sustainable environment in our lifetimes. That our contributors come from a range of academic and professional backgrounds is testament to the ecumenical approach needed for effective environmental planning into the twenty-first century.

Susan Buckingham-Hatfield
Bob Evans

1

Achieving Sustainability through Environmental Planning

SUSAN BUCKINGHAM-HATFIELD and BOB EVANS

It would not be unreasonable to argue that Britain was the birthplace of modern land use planning. To be sure, there were many earlier instances of town design or even environmental planning – but it was in Britain, in the first half of the twentieth century, that the principles of what was to become a national system of land use control and planning were first established. The early reforming pioneers of environmental management and town design – Howard, Geddes, Sharp, Abercrombie, Peplar and others – through their energy, vision and charisma, were to play a major part in establishing British town planning, firstly as a social movement intent on reform, and subsequently as a profession and an activity of government.

The early Garden Cities and the post-war New Towns; the principles of urban containment through green belts; the designation of the National Parks; the organisation of city reconstruction and redevelopment through Comprehensive Development Areas – these and other initiatives were widely regarded by contemporary observers as positive and radical developments which would permit Britain to slough off the legacy of the industrial revolution and move forward to an era of rational planning, greater social equality and widespread prosperity.

Environmental Planning and Sustainability. Edited by S. Buckingham-Hatfield and B. Evans.
© 1996 by John Wiley & Sons Ltd.

Looking back on these years from the vantage point of the last decade of the twentieth century, it is easy to see the fault lines, the pretensions and the mistakes, and equally, it is possible to understand and explain these in the context of Britain's post-war reformism and subsequent 'long boom'. Nevertheless, during the two or three decades after the ending of the Second World War, 'town planning' as a profession and an activity of government became established as the peculiarly British mechanism for managing 'the environment'.

However, until comparatively recently, 'the environment' has had little or no real currency in policy terms. On the contrary, during the 1950s, 1960s and 1970s, politicians and policy-makers largely assumed a position of environmental neutrality: the environment was there to be utilised, sometimes exploited, sometimes conserved, but always in the interests of 'society'. The idea that there were questions of limits to either growth or environmental exploitation were notions confined to academics and campaigners.

Perhaps the most important single publication to change all this was the Brundtland Report (WCED 1987). Many writers have argued that *Silent Spring* (Carson 1962) or *The Limits to Growth* (Meadows *et al.* 1972) were of pivotal importance in transforming attitudes towards the environment, and this is no doubt correct, but it was Brundtland that placed the notion of sustainability firmly and immovably on the policy agenda, internationally, nationally and locally. The agreements that emerged from the 1992 Rio 'Earth Summit' and the development of current European Union environmental policy (CEC 1992) are both substantially a reflection of Brundtland, and as a consequence of these two, sustainability, or at least sustainable development, has become a central organising concept for environmental policy in the 1990s.

However, it is clear that sustainability and sustainable development are contested concepts. Despite the widespread usage of these terms in the context of public policy at all levels of government, it has to be recognised that there is currently a wide range of interpretations as to their meaning, and furthermore, it is equally clear that any significant level of agreement is unlikely to be achieved in the short term. We will argue below that this lack of definitional clarity and unanimity of purpose should not discredit sustainability as a political goal and policy objective – on the contrary, the fostering of a lively and informed public debate is likely ultimately to benefit the move to a more sustainable world – since we maintain that sustainability is, at its very heart, a political rather than a technical construct.

If environmental sustainability is the policy goal, however defined, environmental planning is the mechanism for getting us there, and as with sustainability, there is, in Britain at least, a range of views and interpretations as to what this mechanism should be. In this sense, the central themes of this book – environmental planning and sustainability – are integrally connected and the focus of political contestation, and thus our first task is to examine these two themes in turn in order to provide a starting point for the discussions in subsequent chapters.

The problem of sustainability

In Britain, sustainable development has now been formally established as a policy goal at national and local levels. Moreover, the British government is a signatory to international agreements, global and European, which specifically commit it to the pursuit of sustainable development. Furthermore, both the major UK opposition parties have declared their commitment to achieving sustainable development. However, as Myerson and Rydin point out in Chapter 2, no one is opposed to sustainability: it is a concept that has the capacity to span a wide range of environmental positions and social and economic interests. It is the very ambiguity of the term that makes it so attractive.

Clearly, much has been written on the definition of sustainability, and in particular on the distinctions to be made between 'sustainability' and 'sustainable development' (Jacobs and Stott 1992) and the associated distinctions between 'ecologism' and 'environmentalism' or between 'dark green' and 'light green' perspectives on the environment (Dobson 1990). It is neither necessary nor appropriate to rehearse these here. We regard it as axiomatic that sustainability and sustainable development are contested concepts. What we wish to investigate is the implications of this for the policy process, given the supposed centrality of these notions for environmental planning and policy in contemporary Britain.

This leads to our first point, which is that the policy goal of sustainability is, in our view, distinctive and qualitatively different from other policy goals traditionally associated with UK central and local government. Such policy goals are usually *transitional* and *specific* in that they aim to take a society from A to B (from high inflation to low inflation, from public ownership to private ownership) over a specific, and normally short period of time. In contrast, sustainability is in principle both *long-term* and *all-embracing*. It is an implicitly open-ended commitment which is not necessarily compatible with party political goals focused on the short term.

Given this, it is clear that sustainability (or sustainable development) is a very different policy goal from, say, 'full employment', 'more social housing' or 'lower public expenditure'. These are the regular stuff of party politics and manifestos and they are, at least in superficial terms, easily understandable and allegedly quantifiable. In principle, you know when you have achieved the policy goal. In contrast, sustainability is not obviously any of these things.

We therefore suggest that the policy goal of sustainability can be usefully understood as what might be termed an 'overarching societal value'. In this sense, it is more akin to notions like 'freedom', 'justice' or 'democracy' than to specific policy or manifesto commitments. In many, if not most societies there is likely to be either an explicit or implicit acceptance of such overarching societal values, although, of course, it must be recognised that in most societies there is likely be a major chasm between the rhetoric and the reality. Moreover, as with the case of sustainability, there are likely to be many competing interpretations of what any particular value might represent and how it might be achieved.

Our second point flows from this. In our view, sustainability is, at its very heart, a political rather than a technical or scientific notion. The concept represents a *belief* in the absolute necessity for current generations to act as stewards of the earth for future generations, a belief which is fuelled by technical and scientific evidence but which is not determined by this. As Blowers (1993) and others have argued, there are severe limitations to scientific knowledge in this area. Precise cause and effect is often neither clear nor conclusive, the forecasting of future impacts is problematic, and data and scientific evidence are liable to misrepresentation and manipulation by powerful vested interests. The goals of environmental policy, including sustainability, are overwhelmingly political. These goals are informed, or confused, by scientific or technical evidence, but the process of identifying and establishing the goals is political and conflictual, and quite rightly so.

The corollary of this is, of course, that any attempt to 'technicise' sustainability is doomed to failure. There can be no doubt that the current interest in 'Sustainability Indicators' will provide a useful and thought-provoking series of standards against which policy action might be judged. However, it has to be recognised that these indicators cannot be a proxy for sustainability, and achievement of them does not equate with the achievement of sustainability. It is very possible that the state of sustainability – a kind of environmental stasis – will never be reached, and for this reason, it is very much more useful to understand sustainability as an organising and guiding principle in decision-making which may alter in character, degree or emphasis as time and circumstances change (Agyeman and Evans 1994).

Our third point is that the debate over the new environmental agenda and the concept of sustainability is an exclusive debate. It is the preserve of a comparatively small group of people – academics, politicians, activists, administrators – the majority of whom are members of the educated and articulate property-owning middle class. In contrast, the overwhelming majority of the population, in the UK and elsewhere, has been largely untouched by environmental concerns. For most 'ordinary people', questions of crime, poverty, unemployment and other major concerns clearly overshadow environmental worries. There is not much evidence of an 'environmental culture' (O'Brien 1993) developing beyond an elite social grouping, and where it does, it is much more likely to be a concern with environmental *regulation* – of sewage outfalls, of industrial emissions – than with a wider concept of sustainability. The exclusivity of this debate is a cause for concern for two reasons. First, it is becoming apparent that 'the environment', like 'poverty', is fast becoming a source of income and status for a whole army of experts. This is in itself not surprising and perhaps not noteworthy, but there is a possibility that the round of conferences and the weight of publications of recent years may represent a real concern which may not directly translate into action.

Secondly, as all the post-Rio documents make clear, two central elements of sustainability are the notions of equity and democratisation. The policy documents and literature surrounding the Agenda 21 process at local, national and international levels make it very clear that sustainability can only be secured if there is widespread popular involvement in the process, and if there is to be some measure of equality of outcome in a more sustainable society. As a consequence, the Local Agenda 21 process in particular places a heavy emphasis upon citizen participation and involvement, and upon the less clearly specified but nevertheless central concepts of empowerment and capacity building, issues which are developed in more detail below.

The concern here is that the 'process' may become the policy. The Local Agenda 21 process may fall prey to becoming an exercise in consultation and public discussion, with this becoming the central aim, to the exclusion of other, more tangible policy outcomes. A tendency to focus upon process rather than outcomes is perhaps unsurprising since, in most cases, the policy options that may emerge from consultation and participation with citizens and local organisations may simply be undeliverable within existing social, economic and political constraints.

Our final comment on sustainability concerns its acceptance. Even in its mildest, environmentalist, light green form, sustainability involves a challenge to the established order, and in particular it implies the widespread adoption of new patterns of living. More specifically, in the relatively prosperous countries of the North, it is likely to require the acceptance of policies that will threaten current life-styles and patterns of consumption, for example by restricting car usage and mobility, or by reducing levels of consumption and consumer choice. Given that the principal rationale for this is that it is to benefit as yet unborn generations, it is a call for individuals to vote against their own objective short-term interests. This kind of social altruism does not come easily, nor, in societal terms, cheaply. The case for community environmental education or education for sustainability has been made elsewhere (UNEP-UK 1992; Agyeman and Evans 1994), but its importance is unarguable. It seems very likely that the immense changes in public attitudes which are a precondition for the acceptance of policies for sustainability will only occur after a prolonged and purposeful campaign of public information and debate.

Environmental planning

It is commonplace for academics, journalists, politicians and others to refer to 'the planners', in a general and undifferentiated way, meaning that group of expert advisers who appear to have responsibility for the environment. Our purpose in examining the notion of environmental planning is, therefore, to consider the various interpretations of this concept and, by implication, also to review the role and status of this group of people widely designated as 'the planners'.

The term 'environmental planning' may be broadly understood to refer to the process of formulating, evaluating and implementing environmental policy. However, this simple definition serves to conceal a variety of interpretations of environmental planning which may be variously conceived as the *technocratic, professional* and *political* views of the process. There are, of course, overlaps between each of these interpretations, but each represents a distinctive way of looking at the processes of environmental policy and management.

Technocratic interpretations of environmental planning refer to an understanding of the process which gives precedence to the scientific knowledge deemed necessary to understand and thus manipulate natural biophysical and ecological systems. The approach does not deny the importance of the socio-economic environment within which these processes might be managed or planned, but it does tend to emphasise the importance of technical approaches and solutions. Selman's (1992) book on environmental planning exhibits many of these features. As might be expected, this approach places greater emphasis upon rural and natural environments than upon urban, built environments.

Our second category of *professional interpretations* encapsulates a rather different approach to environmental planning. Here we are referring to the responses of established professional groupings to the rapidly expanding environmental agenda and the associated sources of employment and prestige. In particular, the Royal Town Planning Institute, representing the interests of land use planners, has historically claimed a particularly wide environmental remit, often extending well beyond the regulation of land use and the control of development. Although some other professions, for example environmental health officers, are currently seeking to be known as the pre-eminent 'environmental' profession, it is the town planning profession which has been most effective in this respect in its search to move from town planning as its cognitive base, to a more ambitious notion of 'environmental planning' incorporating a much wider policy remit.

Various aspects of this phenomenon have been examined recently by writers such as Healey and Shaw (1994), Owens (1994), and by Marshall in this volume (Chapter 8), but it is by no means clear that a profession based on land use and a tradition of town design has the epistemological competence – the necessary skill and knowledge base – to justify its claims. As has been argued elsewhere, it is unlikely that any one occupational group will be able to legitimately claim the breadth of wisdom and experience, and the theoretical knowledge necessary to oversee the whole of the policy field implied in the designation 'environmental planning' (Evans 1995).

The final view of environmental planning that we wish to consider is the *political interpretation*. Here, the long-term objective of sustainability is linked to a policy process of environmental planning. The clearest expression of this position is that articulated by the Town and Country Planning Association (Blowers 1993), who argue that the only way of dealing with the complexity of environmental questions is to adopt an integrative, holistic policy approach which transcends established departmental and professional boundaries. This interpretation of environmental planning sees it as a set of arrangements for formulating, organising and delivering policy with the objective of securing environmental sustainability.

The Town and Country Planning Association is a long-established pressure group with its roots in the garden city movement inspired by Ebenezer Howard, and as such it has a long association with the British tradition of town planning and town design. However, the TCPA's call for environmental planning is one that places questions of land use within a wider context, incorporating a whole range of environmental policy including pollution control, waste reduction, reclamation and reuse, energy management, transportation planning, water resource management and so on. In this context, the traditional concerns of land use planning are but one part of environmental planning.

Clearly, the kind of environmental planning envisaged by the TCPA does not yet exist, at least in Britain, although the Dutch system of environmental plans is in many ways very similar. Thus, the TCPA is involved in a political, campaigning agenda which seeks to convince government and opinion formers of the necessity to create a system of environmental planning linked to some vision of sustainability. However, in our view, this third interpretation of environmental planning, as a policy process linked to sustainability, is the most useful perspective. Not only does it recognise the importance of creating policy and decision-making approaches which reflect the complexity of environmental questions, but it also implicitly emphasises the political and conflictual nature of the drive to sustainability. Conversely, the approach rejects the professionalised position that environmental questions are the exclusive preserve of credentialised experts.

We will investigate some of the implications of this approach to environmental planning below, and in particular we will refer to the need for strategic and regional mechanisms and plans to secure implementation, and to the difficult but nevertheless crucial links that must be made between the process of environmental planning, the goal of sustainability, and questions of equity and democratisation.

Environmental sustainability and democracy

We have argued above that environmental sustainability cannot be accepted uncontested. Indeed, as Myerson and Rydin propose in Chapter 2, if sustainability is to be a robust and workable goal, it must be subject to debate. Further, we would maintain that this is but the first step in a process of inclusion through which groups and individuals will come to own and take responsibility for planning for environmental sustainability. The rhetoric surrounding these concepts embraces not only environmental goals but also social goals such as greater equality, citizen empowerment and active public participation in environmental decision-making which both the United Nations and the European Commission's Fifth Environmental Programme regard as integral to environmental planning. This is not to say that such social objectives are required to achieve a sustainable environment as, for example, Pepper (1993) warns, but this book argues that the 'sustainability' project should incorporate greater levels of participatory democracy than is currently the case in environmental policy-making.

As public participation in land use planning has consistently demonstrated, making provision for participation is not sufficient to ensure that it will take place in a fair and even manner and examples of selective representation abound. (See, for example, Kirby (1981) and Lowe and Goyder (1983) for a general discussion, and Greed (1994) for the specific constraints on women.) The justification for greater public participation is twofold: social justice on the one hand and functional legitimation on the other. If people feel that they 'own' the decisions made, then they are more likely to want to comply with them.

Nevertheless, as Hirsch (1981) explains, the increasing centralisation and bureaucratisation of power not only erodes the capacity of local government to act independently, but this erosion of local autonomy also results in an increasing disinclination for people to support traditional political groups which, they feel, are unable to influence centralised authority. This, he argues, has led to a rise in grass roots groups which have a tendency to be both more heterogeneous and more decentralised than political parties. However, this very heterogeneity and decentralisation, together with their relatively unstable organisation, makes these groups vulnerable to being 'picked off' by governments or to having one group's interests played off against another's. Ultimately, opposition, or campaigning, of this sort is quite fragile and arguably could lead to a less democratic form of representation. Awareness of this is critical, since Agenda 21 advocates a form of participation that is not necessarily channelled through local government or presided over by it exclusively.

As Judith Matthews demonstrates in Chapter 3, there is potential conflict between social and environmental objectives which can undermine attempts to move towards environmental sustainability. However, this conflict is by no means axiomatic and might be averted if people are invited to have a real and informed say in making the decisions. As Myerson and Rydin point out in Chapter 2, extending the debate on sustainability to local communities will in itself contribute to a general understanding of sustainability and sustainable development.

The Rio Declaration on Environment and Development embodies elements of this in its Principles: the eradication of poverty (Principle 5); the participation of all concerned citizens in the setting of the environmental agenda (Principle 10) whilst ensuring a right of access to information (Principle 11), education, redress and remedy; and the full participation of women (Principle 20), youth (Principle 21) and indigenous people and their communities and local communities (Principle 22). Accordingly, signatories to the Declaration are required and expected to build these principles into their plans for achieving environmentally sustainable development in the twenty-first century. (For an interpretation of 'development' which informs this book, see Blowers 1993.)

The mechanisms for achieving effective participation, however, remain resolutely under-explicit and it is our contention that conventional ways of formulating policy will be insufficient to the task of achieving environmental sustainability. These conventions (such as top-down planning and a method of public participation that is reactive and constrained by the existing decision-making process) need to be challenged at a number of levels. The principle of subsidiarity embodied in United Nations and European Union legislation requires that environmental decision-making needs to be devolved and makes specific reference to the involvement of individuals and groups hitherto neglected in the environmental decision-making process. Contemporary ways (particularly in the United Kingdom) of soliciting public participation have repeatedly failed to involve the groups included in the above Principles, and consequently, it is likely that this form of public participation will need to be recast if it is to involve a wider range of people.

Current transnational environmental initiatives see 'the community' as an appropriate unit of certain elements of environmental decision-making as well as a vehicle for participation. There are several potential problems associated with this: first, the greater involvement of 'the community' as participants in, or as a focus for, environmental planning will not, in itself, lead to a more environmentally sustainable future. Secondly, 'community' needs to be defined; it is being increasingly challenged as a planning concept (Evans 1994) and there is evidence that associations of people not conventionally seen as 'communities' may be more appropriate units with which people identify themselves and in which they are more likely to engage in environmental action. Such units may be the workplace, parent–teacher associations or single-issue associations formed, for example, to oppose a road-building scheme. Indeed, it is to such units that some local authorities are turning in an attempt to reach 'non-joiners' and those people who would not typically become involved in a public inquiry or forum. Whilst this may overcome the problem of representation, the issue of social conflict within and between groups will still need to be resolved. It also reaffirms the potential problem, referred to above, that these groups may be weakly organised, fragile and vulnerable to manipulation.

Furthermore, 'social justice' and 'community' are not necessarily inherently compatible concepts: 'community' can be seen as a constraint on the individual and there is a debate surrounding 'citizenship' which concerns a loosening of tradition and community (Steward 1991), the proponents of communitarianism notwithstanding. Thus, the dual project of social justice and environmental justice is significantly more problematic and in need of resolution than an initial reading of United Nations and European Union principles of citizen involvement may suggest.

There are other aspects of the sustainability approach which also need problematising. The United Nations, by highlighting co-operation and participation by disadvantaged people, brings to the policy debate the issue of which and whose knowledge should be privileged. If women's, young and indigenous people's views are to be actively sought, are they going to be actively incorporated into environmental policy itself and/or the environmental policy-making process? Moreover, the inclusion of these groups is simply the starting point for involving hitherto voiceless groups: there is little acknowledgement in the environmental literature of the difference within these groups and this, too, will need to be addressed in developing strategies for environmental sustainability. (For the implications that such differences may have when women's views on environmental issues are sought, see Buckingham-Hatfield 1994.)

Current consultation practice has led to a degree of cynicism about the efficacy of participation amongst people and groups who feel disenfranchised. Policy-makers will need to ensure that the environmental policy which emerges out of consultation and Agenda 21 represents the input of all participants and is not simply 'mediated away' by the lowest common denominators of political and economic expediency. If this cannot be achieved, the cynicism which now surrounds the participation process will erode people's ability to 'own' and take responsibility for an environmentally sustainable future. To ensure that a policy fairly represents the participation process requires a degree of management and intervention which, as Matthews suggests in this volume, can itself generate conflict which may not easily be willed away by using the 'right' kind of participation or democratic accountability. So, participation, too, is both a problematic and a political concept which, along with sustainability, needs exhaustive debate to ensure that it is properly and effectively understood and deployed.

Situating the environmental debate

The environmental debate is currently framed by a 'way of knowing' that is 'Western', 'scientific' and 'rational'. For several of this book's contributors this is a major obstacle in the pursuit of environmental sustainability. Indeed, ethnic minorities in Britain, the 'third world within', do not necessarily share the romantic view of nature which Agyeman and Tuxworth propose in Chapter 7 as part of the heritage of environmental concern in the United Kingdom. Bayliss and Walker further argue in Chapter 6 that science itself is used spuriously in that it is impossible to measure and quantify everything and that environmental decisions must therefore be based on partial knowledge rather than the 'absolute' or 'objective' knowledge frequently claimed by scientists. (See, for example, the resurfacing of the debate on the partiality or impartiality of science rehearsed by Wolpert and Collins 1994 as quoted in Irwin 1994.) The 'measurement' of environmental degradation by economic criteria (as in the 'polluter pays' concept embraced by Western governments in their search for an acceptable environmental policy) may also be questioned in so far as economic cost and/or benefit is seen as a constraint on minimising such degradation.

Von Weisacker (1994) argues that the primacy of economics needs to be overwritten by the environment if we, as a society, are to achieve environmental sustainability in the twenty-first century. He argues that 'economics', as the defining concept of the twentieth century (as the 'nation state' was for the nineteenth), is not adequate to the task of tackling environmental degradation on the scale now witnessed, and that the new millennium needs an alternative defining concept which he proposes should be environmental. Porter and Welsh Brown (1991) hold that the world has already embraced a new paradigm, with 'sustainable development' replacing 'neo-classical economics' as a framework for understanding and conceptualising the environment since the late 1980s, although several of this book's contributors would contest that this transition has been accomplished. In this vein, Rydin and Myerson also argue that 'sustainability' could be the post-modern equivalent of a grand narrative which replaces the modernist grand narrative of progress that has informed the twentieth century.

Mellor, in Chapter 4, suggests one alternative way of understanding society/environment relations that may explain environmental degradation and which can offer radical strategies for achieving a more environmentally sustainable society by ceasing to privilege the Western 'scientific' approach that has underlain the exploitation of the earth. McLaren follows a similar argument in Chapter 9 in stating that hierarchical social relations (between 'races') are responsible for environmental degradation which, in addition, has a disproportionately negative effect on disadvantaged people.

If, then, we are to follow the United Nations and European Union injunctions to engage all people in the environmental debate, are we to do so within existing social frameworks and run the risk of continuing to exclude people from the real decision-making, or do the terms of the debate and, therefore, the structure of decision-making, need to be altered to incorporate voices hitherto neglected? Agyeman and Tuxworth argue in Chapter 7 that, despite structural constraints, local government is moving towards greater environmental sustainability through engaging with some of these issues, despite the constraints imposed upon it by the UK central government. Nevertheless, this is presenting an enormous challenge to local authorities in the UK – diminished in capacity and authority by years of centralisation of power – to re-conceptualise methods of policy development and the role of government. Both Agyeman and Tuxworth (Chapter 7) and Marshall (Chapter 8) argue that the European Union, post-Maastricht, through greater emphasis on the environment, the region and the elected European Parliament, is enabling a dialogue between the local and the global that may, in the not too distant future, transform environmental planning. That transformation is part of wider political changes anticipated by Hobsbawm (1991) and von Weizsacker (1994) in which the nation state has diminishing importance.

This book, then, is a multidisciplinary attempt to demonstrate how sustainability and sustainable development can be achieved by a form of environmental planning that incorporates concepts of equity and democracy. It calls for a widening both of the debate and of responsibility-sharing for the environment so that this is not seen as the preserve of a professional elite of elected and appointed officials.

The book's structure

The following chapters can be broadly divided into two sections, with Myerson and Rydin's (Chapter 2) and McLaren's (Chapter 9) forming the internal anchorages. Matthews (Chapter 3), Mellor (Chapter 4) and Agyeman and Evans (Chapter 5) address issues of environmental inequity in that they consider how local people, women and people of colour have both borne the bulk of environmental externalities and are likely to pay disproportionately for strategies to ameliorate these unless reforming steps are taken. Bayliss and Walker (Chapter 6), Agyeman and Tuxworth (Chapter 7) and Marshall (Chapter 8) all consider the specific role of policy and planning in the UK (although a UK which is part of the European Union) and how sustainability may or may not be achieved through prevailing strategies.

George Myerson and Yvonne Rydin open the debate in Chapter 2 where they explore the political nature of sustainability and sustainable development through considering the nature of the discourse that defines these terms and through which political agreement may be reached. They argue that a process of argumentation by which all parties can articulate and learn more about sustainability is critical, and needs to be iterative, if we as a society are to plan for sustainability. At present they perceive the planning profession to be ambivalent about its responsibilities towards the environment and economic development.

A parallel point is made by Judith Matthews who argues in Chapter 3 that, whilst the planning profession is busy attempting to 'capture' the new and intellectually lucrative expert area of environmental planning, it is hampered by its role as a mediator of conflict. A common conception amongst 'non-experts' is that investment for a more sustainable environment must necessitate a reduction in investment elsewhere, such as housing, welfare or health. To overcome this, Matthews believes that planners need to pay greater attention to understanding the relationship between people and the places in which they live and work, so that by reflecting on the social impacts of environmental change, planners can mitigate the harsher effects of environmental planning. Planners should also take their share of responsibility for educating people to the idea that whereas a move to a sustainable environment will require life-style changes, as McLaren later points out, these do not automatically translate into a reduction in well-being.

In Chapter 4, Mary Mellor distinguishes between those who currently bear the brunt of environmental externalities and those who, through the material reality of others' lives, live as though environmental limits do not exist. Whilst, as she says, human beings are embodied (that is constrained by their physical needs and their mortality) and embedded (in their ecosystem), Mellor argues that women disproportionately bear the burden of this embodiedness and embeddedness because of the way in which society values their experience. Such an analysis leads her to the conclusion that a more fundamental shift in practice and attitudes, which values women's experience, is required if a sustainable environment is to be achieved.

Whilst Mellor includes colonised people as sharing this undervaluing by a culture dominated by white maleness, this theme is taken up explicitly by Agyeman and Evans in Chapter 5. They examine the link between environmental degradation and economic power in a racialised world, where racism and economic exploitation are inextricably intertwined with environmental futures. They argue that global and local environmental debates cannot be fully comprehended unless the crucial issues of race and ethnicity are adequately recognised and understood.

The following three chapters relate explicitly to the planning structure in Britain and the European Union and question the ability of planning as currently constructed to achieve sustainability. Through an analysis of Environmental Impact Assessments and environmental monitoring procedures, Duncan Bayliss and Gordon Walker (Chapter 6) argue that a lack of adequate and comparable information seriously hampers government's ability to set targets for the attainment of sustainability. Not only that, but the tendency to privilege technical and scientific knowledge (and hence technologists and scientists) prevents us from taking precautionary action and a longer-term perspective – a point later stressed by McLaren.

Julian Agyeman and Ben Tuxworth (Chapter 7) focus their criticism on central government which, they believe, has abandoned any inclination to develop a proactive strategy. What environmental policies it has espoused have largely been in reaction to international (global or EU) agreements. On the other hand, the most active level of government with regard to the environment has been local government. Agyeman and Tuxworth see that developments in Europe which encourage direct links between local areas in member countries and between regions are likely to be positive with regard to pursuing environmental sustainability.

Tim Marshall (Chapter 8) explicitly explores the role of the European Union in British planning and the importance of various European networks in developing environmental programmes. He is more positive about the heritage of British land use planning in setting the environmental agenda in the UK, but shares with Agyeman and Tuxworth the recognition of the importance of non-governmental organisations (NGOs), particularly the increasingly positive relationship between NGOs and local government.

Duncan McLaren (Chapter 9), himself representing one of these NGOs, draws together themes developed throughout the book in a concluding chapter which argues that 'sustainability planning' is essential and can only be achieved through the setting of clear and effective targets that are widely publicised. He feels that government has for too long favoured a minority (namely, corporate interests) through requiring environmental interests to compete with economic ones, rather than act as their constraint. McLaren emphasises one of the book's key themes, democracy, arguing that people need accurate and comparable data so that they can make well-informed decisions regarding their behaviour towards the environment. Moreover, he suggests that government should create a legislative climate in which these decisions can be effectively implemented. He points out that the constraints to sustainability planning are political and ideological, emphasising the political nature of sustainability articulated at the beginning of this introduction.

Our contributors, then, though focusing on different aspects of planning for a sustainable environment, and coming from a range of backgrounds, are united in their call for a sustainable future that responds to all peoples' needs and democratically expressed desires. Such desires can only effectively be expressed if there is wider, free access to information and to environmental education. Whilst environmental decision-making needs itself to be democratised, the contributors see a role for planners, and each offers ways in which they feel sustainability might be achieved.

2

Sustainable Development: The Implications of the Global Debate for Land Use Planning

GEORGE MYERSON and **YVONNE RYDIN**

> Sustainable development is difficult to define. But the goal of sustainable development can guide future policy. (Foreword by J. Major to *Sustainable Development – The UK Strategy*; HMG 1994a.)

In 1991 the Department of the Environment issued a guide for government departments entitled *Policy Appraisal and the Environment*. This slim volume is presented with a green cover, of course, based around a picture of spring sunlight filtering through new beech leaves. It provides a simple introduction to the principles of environmental economics, as recommended in the Pearce Report (Pearce *et al.* 1989) *Blueprint for a Green Economy* (itself a result of the government's concern over the possible implications of the concept of sustainable development). In the introduction to the DOE guide is the following quotation:

> To White Hall, where met by Sir W Batten and Lord Brouncker, to attend the King and Duke of York at the Cabinet; but nobody had determined what to speak of, but only in general to ask for money. So I was forced immediately to prepare in my mind a method of discoursing. (*The Diary of Samuel Pepys*, 7th October 1666 in DOE 1991)

Here we find the juxtaposition of the two key themes of this chapter: the incorporation of new environmental concerns, sustainable

Environmental Planning and Sustainability. Edited by S. Buckingham-Hatfield and B. Evans.
© 1996 by John Wiley & Sons Ltd.

development concerns into policy; and the recognition that the policy process is a matter of 'discoursing', of communication, of arguing.

The purpose of this chapter is to explore the terms 'sustainability' and 'sustainable development' as they are used in various policy statements, including those outlining a role for land use planning. In doing so, an approach is adopted based on the recognition of policy as argumentation. The role of argument in policy development and indeed of policy as argument has been discussed by various authors such as Braybrooke (1974), Lindblom and Cohen (1979), Majone (1989), Hood and Jackson (1991) and Fischer and Forester (1993), and is in line with the general post-positivist shift in the social sciences, the discovery of the linguistic or (preferably) argumentative turn and the greater integration of concepts from the humanities and social sciences.

The idea is to study policy not simply in terms of a process of policy formulation, statement and implementation, but rather to see all stages as both interconnected and constituted through communication between actors. Communication between actors shapes policy, gives expression to the constraints on policy practice and itself provides an additional constraint since only that which can be communicated can enter into the policy process. The analysis of the way in which communicative acts influence policy is revealing about both the detailed operation of that process and the ways in which factors such as conflicts of economic interest feed into policy. To understand fully the expressions of conflict and the creation of consensus at the heart of the policy process, the emphasis on communication as argument, on a rhetorical approach to discourse is particularly helpful. This approach has been used by the authors in relation to green belt policy (Rydin and Myerson 1989), the use of the term 'environment' in planning documents (Myerson and Rydin 1994) and the operation of environmental policy at the urban level (Ave *et al.* 1994). It should be emphasised that this use of the term 'rhetorical' refers to the development of classical analyses of argument, and not the derogatory description of particular speeches.

Here the rhetorical approach is used to consider the implications of the terms 'sustainable' and 'sustainable development'. A variety of surveys have shown an increasing concern with these concepts within the planning system (Marshall 1992b; Morgan *et al.* 1993; Healey and Shaw 1994; Myerson and Rydin 1994; Owens 1994). Commentators have frequently pointed to uncertainties and ambiguities in the use of these terms. This indeterminacy can be seen as a disadvantage, a problem. From a logical perspective, ambiguity can be viewed as providing evidence of muddled thinking on sustainability issues and of the way in which the planning system is avoiding some of the difficult choices implicit in the adoption of a sustainability goal. On this analysis, the term is adopted without a clear purpose. From a practitioner's perspective, uncertainty is problematic in developing key policy guidelines and procedures for achieving this goal.

For example, recent research into the changing local authority mechanisms for regulating the supply of land for housing found that:

> Few authorities refer to sustainability in relation to housing development as a separate issue. Many consider that environmental principles are already embodied in their plans. However, the concept of sustainability is being interpreted in different ways, mainly in support of existing planning policy stances. Despite this emphasis on the environment, local authorities feel unable to develop specific policies on sustainability until the government produces clearer guidelines. There is a general lack of understanding of what sustainable development involves and what should be done about it in practice. (Joseph Rowntree Foundation Findings: *Housing Research* 110, March 1994.)

This lack of definition is presented as a problem for local planners.

In this chapter, by contrast, a more positive approach to such ambiguities and uncertainties is adopted. In some contexts, ambiguity is potentially a creative resource, though it can also carry risks. From an argumentative perspective, the planning system and its policy processes are sets of communicative acts, with actors engaging in argument, debate and persuasion *vis-à-vis* each other. In these acts, they use various resources offered by linguistic and other cultural systems of representation (pictorial imagery or body language for example); the ambiguity inherent in these systems can be valued as a positive feature by actors in their communication and need not be seen as just a bar to direct understanding of each other's position. Ambiguity is notably a creative resource in written texts, media presentations and inter-personal dialogues. Nor do we refer merely to deceptive or 'tricky' uses of ambiguities. Without ambiguity, there is no scope for debate about the concept of sustainable development. The question is: when is the debate constructive and when is it diversionary – a 'method of discoursing' to quote Mr Pepys.

So this chapter will examine the variety of ways in which concepts related by the theme of sustainability are used in policy documents. We will present interpretations of the related terms: sustainability, development, and sustainable development. The examples analysed are drawn from written texts produced in two different arenas: the supra-national arena represented by the UN, OECD and EU; and the central governmental level which in Britain largely constitutes the Department of the Environment but also includes other national level organisations, both quasi-governmental agencies and NGOs. The analysis is based on close readings of the texts within a rhetorical framework. To preview our own arguments, we make four key points: first, that there is an overwhelming consensus in favour of sustainability; second, that this consensus is pre-emptive in that it avoids some key conflicts; third, that a major source of these conflicts is the emphasis on continued economic development; and fourth, that this potential conflict constrains the way in which land use planning has integrated the concepts.

Our proposal for the future is that what will be needed is not more definitive guidelines for local planning practice but rather an expansion of the local politicised debate on sustainability issues, informed by enriched data and supported through more argumentative resources. We suggest that it is through debatable ambiguity that progress in argument may occur, and not by the elimination of ambiguity, since such an elimination would remove the scope for advancing new arguments. It may be part of the strength of 'sustainable development' as a policy concept that it is open to different interpretations, and, therefore, it becomes a focus for contact between contending positions, both philosophical and practical.

The consensus for sustainability

Sustainability is a term over which there is almost universal consensus, in the sense that no-one employs it pejoratively. It is implausible to present an argument as follows: 'I oppose this pro-posal because it is based on sustainability considerations'. Instead, one must argue by reinterpreting sustainability, perhaps suggest-ing it has been misunderstood or misapplied in context. By con-trast, 'growth' is a concept capable of being applied both affirmatively and pejoratively: it is quite permissible in some con-texts to argue that 'this scheme's emphasis on growth is unacceptable'.

Indeed 'sustainability' could be called the post-modern equivalent of a grand narrative, replacing the modernist grand narrative of progress which held sway for much of the twentieth century. Sus-tainability is our way of seeing the present in the perspective of the future and provides a societal story-line for justifying change. The difference, of course, is that sustainability is a far more ambivalent notion than progress, incorporating far more sense of risk and uncertainty. As we shall further demonstrate below, 'if we do . . . if we do not . . .' is the rhetorical construction that underlies the concept of sustainability. The following provides an example of the rhetorical structure: 'The question however is, do we have to wait until disaster overwhelms us before we make the radical changes necessary to protect our world for future generations? That is the vital challenge of sustainable development. If we act now there is much that can be saved which will otherwise disappear forever' (HMG 1994a).

Sustainability is an exhortatory concept, an appeal to change, in-cluding the menace of disaster, whereas progress is (or, perhaps, was) a confirmatory concept, a demand to 'push ahead'.

And unlike the demand for progress, which was relentlessly op-timistic, there are more and less hopeful variants on the view that achieving sustainability is an essential policy goal. The Department of the Environment's follow-up guide, *Making Markets Work for the Environment* (DOE 1993c; the cover of the publication shows a photograph of autumn leaves in shades of red, green and brown against a black background), states that: 'We have a duty to ensure that our use of the environment is sustainable: we cannot use and degrade it without thought for the future' (p. i), but does not foresee much cost involved in doing so: 'Achievement of environ-mental goals remains a key objective: increased emphasis on the use of economic instruments does not signal a relaxation of environmental commitments. Changing our approach to achieving environmental objectives will however improve value for money.'

Michael Jacobs provides a subtle example of the pessimistic inflection of sustainability, the sense that the best we can achieve is not too much degradation. His definition of the term is taken from *The Green Economy* (Jacobs 1991) and quoted in the Department of the Environment's *The Environmental Appraisal of Development Plans: a Good Practice Guide* (DOE 1993d).

> that the environment should be protected in such a condition and to such a degree that environmental capacities (the ability of the environment to perform its various functions) are maintained over time: at least at levels sufficient to avoid future catastrophe, and at most at levels which give future generations the opportunity to enjoy an equal measure of environmental consumption. (DOE 1993d, p. 1.)

Something strange happens here to the positive and negative alternatives implicit in any use of the term 'sustainability'. 'At least' is the negative pole, the 'if not' pole of sustainability: if it is not adopted, then future catastrophe results. But what about 'at most', the implied positive pole, the 'if' pole of sustainability? The paradox is that normally, 'at most' is a phrase of denial: 'at most we will agree to this much, certainly not more'. It belongs in a context of negotiation, bargaining over claims. The implied bargaining is with future generations, who will 'at most' be offered the present deal, and certainly no improvement. So the alternatives are negative and less negative.

The less hopeful example demonstrates culturally how the use of the term 'sustainability' can be coloured by different emotions though Jacobs treats the possibilities with notable care, personal to himself. However, pessimistic or optimistic, it is generally agreed that sustainability must be a policy goal: 'The long-term objectives of environmental and economic sustainability must be accepted along with that of environmental quality in the short term' (CEC 1990, p. 34).

The key question now becomes how it relates to other prevailing policy goals and, in particular, how the concept interacts with established concepts of 'growth', 'progress' and 'development'.

Sustainability and progress: a pre-emptive consensus?

Our Common Future (WCED 1987) is a key reference point for almost all discussion on these themes. As well as providing the seminal definition of sustainable development, which we will comment on later, it provides other interesting viewpoints on the relationship between sustainability and progress: 'But the Commission's hope for the future is conditional on decisive political action now to begin managing environmental resources to ensure both sustainable human progress and human survival' (WCED 1987, p. 1). In this quote 'sustainable' meets 'progress'. On the one hand, sustainability reinforces progress, making it still conceivable. On the other hand, there is an implied contrast, a contrast with 'unsustainable progress'. It is this negative shadow that makes sustainability such a powerful concept argumentatively. When deployed 'sustainability' sounds affirmative, but it often acts negatively: if we do not achieve sustainable progress, then what passes for 'progress' in the short term will turn out to be unsustainable in the long term. Sustainability offers conditional support to the notion of progress. And survival brings out the element of menace: if not sustainability, then . . . . In effect, sustainability provides progress with a context at once distanced from determinist predictions and resistant to prevailing cynicism. Sustainability is an affirmative sounding way to utter dire warnings that would be too demoralising if phrased in more negative ways.

The next example further demonstrates the consensus; again, there is no question that sustainability is desirable. It also demonstrates the tensions within the consensus. It is taken from the CEC Fifth Environmental Action Programme Towards Sustainability: 'As used in the Programme, the word "sustainable" is intended to reflect a policy and strategy for continued economic and social development without detriment to the environment and the natural resources on the quality of which continued human activity and further development depend.' (Commission of European Communities (1992), Vol. 2, p. 2).

In effect, the term 'sustainable' carries forward the demand for assent to a policy, even while the content of that policy is being redefined; there is the link with development ('continued economic and social development') and then the qualifying negation ('without detriment to the environment'). Significantly, the environment functions as a limit on development, rather than a factor within it. The 'if . . . , if not . . .' logic of sustainability arguments is reflected by 'continued . . . continued . . . further'; if development is not sustainable, then it will not continue. Sustainability is the formula for continuation, and reflects the risk of not continuing development.

These inherent tensions between continuation and change are made even clearer when applied to a particular policy area, transport, and so is the role of sustainability in managing them. Sustainability is now a significant concept in the formulation of transport strategies at international level as in the CEC Community Strategy for 'sustainable mobility' (1992 Green Paper on the impact of transport on the environment). Imagine the following sentence without the term 'sustainable': 'This Green Paper provides an assessment of the overall impact of transport on the environment and presents a Common strategy for "sustainable mobility" which should enable transport to fulfil its economic and social role while containing its harmful effects on the environment' (CEC 1992, p. 5).

A strategy for 'mobility which should . . . fulfil its economic and social role' is more obviously liable to produce 'harmful effects'. The economic and social would be traded off against the environmental, on the basis that they were fundamentally opposed. Sustainability turns the antithesis into a harmony: both sustainable mobility and the economic and social role are contrasted with 'harmful effects'.

This rhetorical strategy can be seen at work elsewhere in the same document: 'A strategy based on a global approach would promote "sustainable mobility" by integrating transport into an overall pattern of sustainable development' (CEC 1992, p. 43). To promote mobility is one goal, to promote environmental regeneration might be another goal: but the antithesis is pre-empted, and integration can be proposed. Again, from the same document:

> The results of this examination should provide the framework for a Common strategy of 'sustainable mobility', which should contain the impact of transport on the environment, while allowing transport to continue to fulfil its economic and social functions, particularly in the context of the Single Market, and thus ensure the long term development of transport in the

Community. It should also contribute to social and economic cohesion in the Community and to the creation of new opportunities for the peripheral regions. (CEC 1992, p. 55.)

Again the social and economic are set against the environmental, as external considerations, potentially opposed. But the use of the term 'sustainable' pre-empts the conflict between social and environmental. The effect is pre-emptive consensus.

The concern with development as a source of tension in the sustainable development debate

So there is an overwhelming consensus in favour of sustainability concerns, a consensus which appears to be pre-emptive in that it avoids potential conflicts between environmental impacts and social and/or economic concerns. The existence of such conflicts is recognised in much environmental literature and has prompted a debate about the appropriate form of development that should be pursued (de la Court 1990; Jacobs 1991; Ekins 1992). This conflict is also reflected in the discussion of development issues and the way that 'development' is juxtaposed with 'environment' in the documents. For development is a concept of even more radical ambiguity than sustainability, and 'sustainable development' is even more complex than 'sustainability'. The cultural advantage is that there is more scope to debate sustainable development than sustainability, and more mobility of application. The cultural problem is: when does ambiguity become dysfunctional?

Starting again from *Our Common Future*: 'These are not separate crises: an environmental crisis, a development crisis, an energy crisis. They are all one.' (WCED 1987). The statement that 'These are not separate crises' carries with it the implication that they might be interpreted (albeit wrongly) as such. The sense of disagreement is more overt than in examples concerning 'sustainability'; the negation more direct, confrontational rather than consensual. This disagreement is also explicit in the *CEC Fifth Environmental Action Programme*: Vol. II: 'All human activity, economic and socio-cultural, either prospers or founders on the quality of the relationship between society and the natural world. Development is only "real" if it improves the quality of life'.

(p. 17). Again, the presentation is more confrontational: the inverted commas around 'real' allude to other views, which might interpret development differently, as a separate goal from environmental policy. There is a more obvious sense that different viewpoints, ideologies, see development differently, than was the case with 'sustainability'.

Even the UK Government's *Sustainable Development: The UK Strategy* (HMG 1994a), which tries very hard to promote a consensual line works by direct disagreement or confrontation with opposed views:

> A society which does not grow is one which cannot satisfy some of our basic human needs. 'Growth is the only evidence of life': Newman's quotation warns us against suggesting that we could promote the shrinking economy as a basis for sustainability. Economic development is just as important a concept as environmental protection, and we must find ways of achieving both together. (HMG 1994a, p. 5.)

Once the term 'development' is foregrounded, then the rhetorical structure is adversarial: 'warns us against'. The underlying point is that not everyone sees the term 'development' as positively as these texts require. It is possible to use 'development' pejoratively. To talk of 'too much development' is culturally plausible, whereas too much sustainability is not culturally plausible. In the next example from *The UK Strategy*, the argument contradicts a hypothetical adversary who denies that conventional growth is an acceptable standard:

> Conventional measures of economic growth do not take into account the impact of that growth on the environment and the loss of some natural resources. Nonetheless, as measured, the UK economy has grown by about a half over the last 20 years, and a number of future projections contain a mid-range assumption that there may be 60% growth or thereabouts over the next 20 years. (HMG 1994a, p. 11.)

In the DOE guide *Making Markets Work for the Environment* (DOE 1993c), the foreword by John Selwyn Gummer (as Secretary of State for the Environment) makes the potential for conflict clear and steadfastly states the potential for overcoming it (in this case the magic wand is provided by the tools of environmental economics): 'Although these objectives might seem to conflict (of environmental policy and economic policy), they do not have to do so. By using appropriate instruments, we can spare industry the straitjacket of detailed regulation and still deliver high environmental standards.' The use of the phrase 'sustainable development' itself encapsulates the points we have been making: conflict being presented and then deflected. The seminal reference is, of course, from *Our Common Future*, itself quoted in many of the other documents we have used in this chapter:

> Humanity has the ability to make development sustainable – to ensure that it meets the needs of the present without compromising the ability of future generations to meet their own needs. The concept of sustainable development does imply limits – not absolute limits but limitations imposed by the present state of technology and social organization on environmental resources and by the ability of the biosphere to absorb the effects of human activities. But technology and social organization can be both managed and improved to make way for a new era of economic growth. (WCED 1987, p. 40.)

Here the ultimate conflict between development and environment is acknowledged but, optimistically, the report discerns a way through the conflict to 'a new era of economic growth'. The concept of sustainable development itself helps articulate the hopeful view: 'The concept of sustainable development provides a framework for the integration of environment policies and development strategies', but the only way to achieve this argument is to fudge the definition of development: 'the term "development" is being used here in its broadest sense' (WCED 1987).

Of course, in many specific policy contexts, the view of development that is adopted is chosen from among the competing alternatives and presents only one possibility. While the UK government feels impelled to argue for the reconciliation of environment and development, rather than taking it for granted, they do insist that their reconciliation is attainable within prevailing development patterns:

> Sustainable development does not mean having less economic development: on the contrary, a healthy economy is better able to generate the resources to meet people's needs, and new investment and environmental protection often go hand in hand. Nor does it mean that every aspect of the present environment should be preserved at all costs. What it requires is that decisions throughout society are taken with proper regard to their environmental impact. (HMG 1994a, p. 7.)

The role for land use planning

Where does land use planning fit into this broader debate about the future, with its presumption of sustainability as a policy goal, its difficult relationship to development debates and the tendency to manoeuvre around disagreements and tensions between environmental and social or economic goals? The UK government has sought to provide an explicit goal for the planning system in achieving the goal of sustainability:

> The planning system, and the preparation of development plans in particular, can contribute to the objectives of ensuring that development and growth are sustainable. The sum total of decisions in the planning field, as elsewhere, should not deny future generations the best of today's environment. This should be expressed through the policies adopted in development planning. (DOE 1992a, para. 1.8.)

and:

> Development plans can offer local authorities a real opportunity to contribute towards the goal of ensuring that development and growth are sustainable. (DOE 1993c, p. i.)

As might be expected, the debate about sustainability and land use planning reflects the structure of the broader debate. There is the assumed consensus for sustainability and the dominance of the 'if . . ., if not' construction:

> Understanding what environmental sustainability requires will not make planning decisions on the ground any easier. Indeed it may make them more difficult as we recognise the limits of the environment's capacity. But this is not a reflection of the objective itself. It is a recognition that sustainable development is necessary: that society must learn to live within its environmental means. If we do not the consequences are clear. If we do – and there is no doubt that we can – the benefits will be equally manifest. (CPRE 1993, p. 47.)

Again the paradox of sustainability also emerges. It works by presenting alternatives: if . . ., if not, and traditionally, in policy deliberation, alternatives imply a decision between possibilities. But these are not the traditional kind of policy alternatives. Really, there is no choice; only a dichotomy between disaster and hope. In land use planning, as elsewhere, sustainability reveals alternatives in order to discount them in favour of one true path.

There is also the problematic relationship to the term 'development':

> Instead of the environment being a residual in planning decisions we should be choosing the environmental standards we need for the future and using planning to achieve them. This means helping the public and politicians to set environmental objectives and identify environmental limits to development where these are needed. It means going beyond the planner's safe haven of 'balancing' development and the environment to identify environmental capacities which cannot be breached. (CPRE 1993.)

Here sustainability is counterposed to more pragmatic approaches to balancing and trading off development and environment.

And there is the tendency to gloss over the depth of the contradiction possible in this relationship and the alternative views on possible development patterns. In PPG 6 on Town Centres and Retail Developments there is the attempt to present a modified version of current retail development patterns as consistent with the goal of sustainable development:

1. The Government's objectives are:
   — to sustain or enhance the vitality and viability of town centres which serve the whole community and in particular provide a focus for retail development where the proximity of competing business facilitates competition from which consumers benefit; and
   — to ensure the availability of a wide range of shopping opportunities to which people have easy access (from the largest superstore to the smallest village shop), and the maintenance of an efficient and innovative retail sector.
2. These objectives are compatible with the aim of encouraging sustainable development, and in particular with Chapter 7 of Agenda 21, the international action programme for development and the environment adopted by the United Nations Conference on Environment and Development in 1992. (DOE, 1993e, paras 1 and 2.)

One way in which the potential tensions have been procedurally resolved in the case of land use planning, is to place great reliance on the environmental assessment of development plans:

> Environmental appraisal in this context enables the environmental consequences of planning actions to be taken into account systematically when development plans are being prepared. Environmental aspirations and impacts can be compared with social and economic ones, and appraisal also enables conflicting environmental issues to be considered. The choices that are highlighted by the appraisal process are made in a policy context. When

that policy context is one of moving towards sustainable development, environmental appraisal will make a positive contribution to that goal. (DOE 1993c, p. 38.)

And again, after reference to *This Common Inheritance*:

Local planning authorities have a key part to play in helping to achieve the vision for Britain and the environment in the 1990s set out in that White Paper. One major responsibility is to ensure that development plans are drawn up in such a way as to take environmental considerations comprehensively and consistently into account. In this way environmental improvement can be plan-led, and individual development decisions taken against an overall strategic framework that reflects environmental considerations. (DOE 1992a, para. 6.1.)

Essentially, environmental assessment is a procedure by which information is collected on the environmental impact of a project, plan, programme or policy with a view to assessing whether the overall impact is acceptable or not and whether the project (or plan, programme or policy) can be altered so that the impacts on the environment are mitigated. The assessment process provides new material for entering into the decision-making process but does not, in itself, ensure a higher level of environmental protection; it depends on the ways in which the issues are debated and the decisions are taken. There is a danger, however, in relying on a procedural device (which is itself not without shortcomings) to achieve the task of resolving a potentially more fundamental tension between environmental and development concerns. The adoption of environmental assessment for projects and development plans may give the impression that the goal of sustainability is being written into the planning system, but the debate over whether planning should primarily seek to facilitate economic development or protect environmental services and assets has yet to be resolved in favour of environmental sustainability.

Conclusions

The analysis of the debate on sustainability and sustainable development has revealed the existence of a pre-emptive consensus in favour of sustainability and persistent conflicts and tensions in the relation between development and environment. This structure of the debate is apparent both at the global level and in relation to the specific debate on the role of land use planning in achieving sustainability. It seems unlikely that a procedural device such as environmental assessment will readily be able to resolve these tensions and conflicts, though it may bring new information and arguments about environmental impact into the planning arena.

As seen in the introduction, there have been calls for more central government guidelines on how planning may help achieve sustainable development, reflecting the incorporation of new concepts into the planning policy process but also recognising (implicitly or explicitly) the problems the planning system faces in trying to deal with these tensions. Our analysis suggests, however, that guidelines are unlikely to help planners in dealing with such conflicts except to provide them with additional strategies for glossing over them. Rather, and following through the logic of an emphasis on the argumentative nature of the policy process for sustainability, it would be more appropriate to open up the policy processes of the local planning system to a broader debate on the relationship between development and environment issues in the local context. Here the ambiguities in the terms and their use in the debate could become a creative resource in the political process, connecting different interests and positions and creating the potential for a less pre-emptive local consensus on sustainability; ambiguity could be constructive, as well as no doubt sometimes diversionary.

Our proposal implies a shift from a concern with sustainability as represented in policy documents informing and guiding the planning system to a concern with debate about sustainability within the planning system, using the various arenas and forums that exist around public consultation, plan preparation and development proposals. To facilitate this shift to debate, there is a need to provide resources in the form of staff, time and (recycled) paper but also in the form of environmental data and analysis. As the good practice guide to Environmental Appraisal of Development Plans says, 'local authorities can only do this [contribute to sustainable development] if they understand the environmental pressures in their area and the impact which planning decisions might have' (DOE 1993b, p. i). We should qualify this statement by emphasising that local communities and local authorities can only add to the debate which might contribute to sustainable development if they have such understanding. Therefore land use planning for sustainability should aim to extend the arguments about the key concepts and their application in a locality, through providing arenas, information and resources for debate. In this way some of the constraints of the global debate on land use planning may be circumvented and more meaning given to the old adage 'think global, act local' in a land use planning context.

3

Social Processes and the Pursuit of Sustainable Urban Development

JUDITH MATTHEWS

Sustainability is a contested concept. In the first place, it has implications that are difficult to define in precise, concrete terms. Secondly, achieving it must inevitably be the result of social action. As Myerson and Rydin have argued in Chapter 2, the way language is used defines different conceptual frameworks, and this may indicate a lack of shared understanding of issues. In relation to 'sustainability' as a concept, there are many different groups which use the term, and the implications of their differing perspectives on what it entails are vital to the evolution of policies and practices for achieving environmental objectives. Urban settings are particularly implicated in this debate since some understandings of the term emphasise the *non*-sustainability of cities. This chapter, then, will discuss some aspects of the dynamics of the social interactions that result from attempts to define sustainability. These interactions contribute to the continually evolving definition of the term, and therefore to the strategies developed in order to attain a sustainable future for cities.

The attribution of meaning to a word or concept does not only represent the interplay of ideological and power positions; it also results from group values. It becomes part of the interaction both

Environmental Planning and Sustainability. Edited by S. Buckingham-Hatfield and B. Evans.
© 1996 by John Wiley & Sons Ltd.

between different groups, and within them. The need to apply criteria of 'sustainability' to the urban environment has led to an upsurge in the level of discussion of strategies by which to improve urban planning in this direction (e.g. Elkin and McLaren 1991; Breheny 1992; Yanarella and Levine 1992a, b; Yiftachel and Hedgcock 1993).

The Town and Country Planning Association working party on sustainability set itself the task of identifying both the political mechanisms and some of the practical land use and management options to promote these objectives. It argued that sustainable development would entail the pursuit of five 'goals' which were defined in the fields of *resource conservation, built environment, environmental quality, social equity* and *political participation*. The report makes a clear and unequivocal claim to contribute to the process of achieving sustainability from the perspective of an 'interested party' – that of a nationally significant British NGO. It has been widely promoted and quoted within the British planning community, and since it incorporates specific recommendations for 'making sustainability happen', it forms a good example of the way in which the 'territory' of a concept is claimed by interested groups. Having been claimed, the definition is then shaped to the values of the group. In the case of the Town and Country Planning Association, as Blowers points out in the introduction to the report, core values focus around the ideals of the 'social city' originally promoted by Ebenezer Howard. The working party's conclusions are consequently 'framed' in terms of these core values, and of a particular interpretation of the way in which Howard's principles should be realised in the late twentieth century (Blowers 1993).

It is now widely recognised, at least amongst environmental practitioners and activists, that sustainability will have profound implications for social and political life. It is also acknowledged that these implications may constitute an important reason for the relatively slow progress toward substantial change in the major objectives laid down for urban plans. This chapter is concerned with the specific issue of the social processes of environmental response, and aims to identify some of the ways in which these processes are likely to affect the success of technical measures introduced to promote sustainability. Various writers have begun to develop 'discourse analyses' of these issues, and in this chapter some of the social processes which give rise to the discourse of sustainable development in cities are directly examined.

The practice of planning has a series of internal and external social dynamics which need to be understood if we are to explain the outcome of certain kinds of action and decision. A more effective theoretical perspective on the nature of these social processes offers the possibility of working with these dynamics rather than being thwarted by their various operations. One framework for such a theoretical perspective is to be found in social psychological work on social representations and social identity (Breakwell and Canter 1993). Social identity theory is concerned with the ways in which membership in social groups explains individual behaviour through the impact of membership on the individual's sense of self (Tajfel 1982). Individuals direct their actions according to their perceptions of the demands of the groups to which they belong. They seek to achieve positive self identity by elevating and promoting the claims of their own group against those of groups to which they do not belong. The work on social identity has generated insights both into the motivations of individuals and into the behaviour of social groups as entities in themselves. Breakwell and Canter argue that there is much to be gained from integrating these ideas with perspectives on 'social representations' – the beliefs, the ways of describing situations or concepts – to which individuals subscribe. Social representations operate at the individual level to allow the person to give meaning to experience by categorizing that experience in ways that are familiar – 'this is how other people like me would understand this person, event, idea'. Representations also serve at the level of the social group as 'public rhetorics used by groups to engender cohesiveness and manoeuvre relative to other groups' (Breakwell and Canter 1993)

Recently, cultural geographers have developed a strong concern with the role of representations in the development of place and space awareness (see for example Duncan and Ley 1993), and with the idea that place is implicated in 'the politics of identity' (Keith and Pile 1993). By integrating social psychological concepts of social identity and social representation it becomes possible to explore how representations are produced and the functions they serve both for individuals and groups. The linking of these processes to issues of space and place suggests a promising approach to environmental planning and the issue of urban sustainability. Explorations of the target of representations, their salience an~~d~~ the relationships between them each reveal aspects both of the processes by which values and attitudes come into being and of the implications these processes have for relationships between groups, and between individuals within groups. In terms of the concept of 'sustainability', we can see that this is a classic case of a term that encodes a 'representation' for many different groups. The social identity perspective alerts us to the mechanisms by which the term has come to hold different meanings for different groups – differences in definition serve to promote distinctiveness between groups. The theory of social representations gives an account of the strong engagement with defining the term amongst its various proponents – representations serve group interests. An integrated perspective between these two areas of concern gives a structure for examining what is being communicated by 'agendas' for sustainability, and also what patterns of interaction are being promoted between and within the social groups that subscribe to it. Since it is a concept that is currently potent both within the formal arenas of environmental planning and management and within an informal arena of public response, the operative processes of negotiation for power are highly significant. Careers as well as communities are in the balance, depending on the outcome of claims to possession of 'the key to sustainability'.

Sustainability and the economic life of cities

One important dimension of the debate over 'sustainability' amongst British and European analysts of planning is the response to the economic crisis of the European city. Whether we consider the massive physical and environmental challenge posed by the renovation of the cities of Eastern Europe, the continuing drive for economic growth of those in Southern Europe, or the problems of restructuring facing many Northern European cities, there can be little doubt that so-called 'traditional' concerns about economic viability and the importance of meeting the needs of the unemployed and the poor in cities will be high on the agenda of urban managers and planners. For urban populations, direct problems of unemployment, poor housing quality and inadequate urban service provision tend to be central to the ways in which they think about what is required of planning for the urban environment.

This does not by any means imply an absence of concern about those aspects of the environment that have come to be labelled 'green'; the high tide of environmental awareness of the late 1980s may have receded somewhat, but it is surely not yet at the ebb. Work at the Norwegian Institute for Urban and Regional Research is illuminating in this respect. Næss (1993) describes a study in which alternative 'scenarios' were developed for five Norwegian cities, based either on a set of seven 'environmental' goals, or upon a projection of current developmental trends. Having evaluated the 'performance' of the cities according to the scenarios based on each condition, interviews were then conducted with two groups, one drawn from politicians and professional planners and bureaucrats, the other from the general population. What was clear was that although both groups agreed on the importance of improving the 'environmental' performance of cities, the general population sample were much less sanguine about the realism of pursuing such goals, and much more concerned than the politicians and bureaucrats about preserving free access to private transport, parking facilities, and allowing development of spacious new industrial sites. Næss concludes that 'The general willingness to prioritize protection at the expense of growth, or environmental regulation over individual freedom, seems to be much greater than the willingness to follow up such priorities through concrete measures in the urban development' (Næss 1993, p. 329). For planners and urban managers, therefore, it remains the case that conflict over planning strategies is virtually endemic, and as the contest between immediate economic interests and more remote environmental considerations intensifies, the planning system faces an increasingly unenviable task of persuasion, to say nothing of the complexities of the technical tasks of promoting sustainable development.

Another dimension of the problem lies in the variations amongst urban authorities in the priorities they accord to different aspects of urban management. A review by Marshall (1992a) gives a useful account of the differences between European countries in the degree to which their urban policies have adopted environmental agendas, and notes the particularly strong conflict of interests between economic growth strategies and environmental strategies that are evident in Southern European cities. This leads him to argue that there is an urgent task to be addressed in selecting the goals of urban strategy, and that such selection can only effectively be made in the light of a better understanding of whether the urban environmental problem is primarily one of form or of scale. The reality that cities are both intrinsically unsustainable (in the sense that they demand a high level of land and resource exploitation, and have little capacity to restore the resources which they utilise) and also possibly our best hope for developing the technologies to promote sustainability (Clayton 1992), makes the task of reducing the level of such damage especially pressing.

It is frequently argued that the more 'sustainable' the city in environmental terms, the better the chance of long-term economic viability, not only for cities but for the global system. Such a perception is, however, far from being generally accepted, particularly if we broaden the scope of our consideration from the British to the European dimension, and even more so if we address the issue globally. The technologies required to reduce environmental damage, limit pollution, and improve the biodiversity of urban systems are expensive and complex, and require different modes of implementation in different settings.

Marvin's (1992) discussion of the potential for using a 'least-cost planning' strategy for implementing sustainability policies is but one example of the kinds of call to action to which planners are subject. It constitutes an attempt to establish a 'representation'; planners face such calls from many quarters, and considerations over the validity, political acceptability, and feasibility of such calls further intensify the problem for planners about how to design and how to implement strategies for change.

Defining the agenda for achieving sustainability

As has already been noted, the recognition that 'sustainability' is a contested concept has come to be commonplace within the literature on this issue (see, for example, Shiva 1993b). Kidd (1992) pointed out that thinking about the relationships between population growth, resource use and pressure on the urban environment has developed at least six different strands since the 1950s. Each is fundamentally different in its assumptions about the future of human life, and each has contributed a thread in the evolution of a concept of sustainability. Because the differences are so fundamental, the resulting prescriptions for a sustainable future are also completely at variance, and cannot be reconciled. Kidd concludes that a single definition of sustainability cannot be achieved. The perspective on social processes outlined above would concur with this conclusion, and account for it as an outcome of intergroup relations. Planners, pressure groups and others concerned with promoting urban change have nevertheless come to identify sustainability as a useful concept to guide strategy. Promoting changes in urban systems is thus increasingly likely to be justified by reference to some version of the 'sustainability' agenda, and variations in the way the term is understood can be expected to form another dimension in debates over the future of urban areas. The suspicion begins to emerge, however, that current prescriptions for sustainability have some of the characteristics of 'premature legitimation' identified by Reade (1987) for the planning system as a whole. The perspective on group processes and representations offers an explanation for this tendency. We can see the rush to claim the territory – whether of 'planning' as a professional skill or of 'sustainability' as an effective concept for advancing planning – as a product of the process of 'representation' by groups wishing to communicate their own legitimacy as agents of environmental management.

Marshall offers a possible way out of this dilemma through his proposition that in practice, the most effective solution to this problem will be for planners to separate and carefully identify goals for social, environmental and economic action leading to sustainability, rather than allowing all the purposes of planning to become subsumed under a single all-embracing category of 'sustainability'. This would appear to be an absolutely necessary practical prescription for action, but there remains the additional task of understanding what the apparently talismanic concept of sustainability can tell us about important conceptual developments within the social system. Analysts of planning need to recognise the structure of the conceptual system implied by the notion of sustainability, if appropriate methods of implementing the goals of this system are to be discovered, and it could be that it is precisely because they have begun to do so that the concept is being developed in the way that it is (see, for example, Whatmore and Boucher 1993).

One of the most familiar aspects of the social dynamics of urban planning lies in the recognition that intervention by planners tends to generate conflict with communities, conflict within communities and conflict between communities. These conflicts are rarely sought by the participants, and indeed are frequently seen as the unintended outcome of a desire to *avoid* conflict. However, the scale of the global environmental threat would certainly seem to make protracted conflict a dangerous option. On the other hand, there is really no reason to suppose that such conflict can simply be willed away or eliminated by adopting 'the right kind' of participation or democratic accountability. Its sources have to be understood before the outcome can be effectively averted. We can also note that some analysts (e.g. Pearce 1992) have argued that the planning system has been effective in *reducing* direct conflict over the use of land. The issue here is really one about the location and nature of the conflict that may have been avoided. Conflict reduction because some contestants have simply abandoned the field is not necessarily a desirable resolution of the problem. As a matter of policy, land use planning decisions seek to mediate between competing uses – and in this sense the system does appear to be effective in anticipating and resolving some conflicts. However, the *consequence* of such decisions will frequently be to generate impacts on individuals and communities which threaten established local conditions, and result in the emergence of new conflicts as groups respond to changed conditions. Thus planning both resolves and generates conflict.

Concern to provide planners with effective mechanisms for developing strategies which will respond effectively to public demand prompted Poulton (1991) to argue the case for a 'positive theory of planning'. He argued that as planning came under political pressure during the 1980s, so the implicit progressive ethos of the system was threatened, and the absence of a clear framework for understanding the purposes and actions of planning became manifest. Planners need to locate an understanding of their role and functions within a framework that acknowledges the demands of the public it serves. The challenges which this presents lie in the fact that time-consuming 'participation' in planning is not a primary requirement for the majority of the urban population, that reaction to change is likely to be couched in terms of immediate self interest on the part of those affected, and that acceptance of any kind of change is likely to be slow in coming. Poulton's proposal for a positive theory of planning would thus seem to demand a significant social dimension if it is to provide, as he advocates: 'a clear line of deduction that is plausible, logically consistent, and unambiguous; a short sequence of steps from the basic idea to testable deductions; and a series of valuable cause–effect predictions that are applicable to significant real world issues' (Poulton 1991, p. 226).

The fact that urban areas face a range of social, economic and environmental problems produces a situation where competing strategies are inevitable, and a further problem for achieving sustainability lies in the fact that the structure of urban governance fragments areas of concern, and produces competing analyses of needs and disparate traditions of action and perception amongst different branches of public and private sector urban authority.

In addition to all this, it is widely recognised that sustainability objectives derived from whichever of the traditions already discussed not only threaten existing approaches, but also constitute a threat to entire current ways of life. Little wonder then that in the face of immediate economic considerations, approaches by both planners and the public should back away from the more risky paths implied by the pursuit of sustainability. Under these conditions, it is highly likely that public support will be weakened by the apparently reduced emphasis on the traditional concerns of urban communities.

The proposals advanced by Healey (1991) are apposite in this respect. She argues the case for a continuing need to develop a theory of planning for practice. She argues that an important element of this will be establishing a better understanding of the nature of the 'communicative practices' of planners if the system is to become more comprehensible and legitimate. For Healey, the pursuit of sustainable development is an important priority for the planning system, requiring better understanding of the processes operating within it. This amounts to a call for attention to the role of representations within the social system. If communicative practices can be identified with the interests of particular groupings within planning, then the potential for adjusting communications to achieve common practical goals comes within sight.

A social process framework for understanding planning for sustainability

As yet, we have only a relatively hazy understanding of the processes by which the concept of sustainability is percolating through the planning system, and very little knowledge of the degree to which the public recognises environmental concerns to have direct implications for urban land use and physical planning. It has been argued above that understanding the processes of urban change can be helped by stepping somewhat outside the normal range of the discussion of planning, and importing insights from the field of social psychology. By understanding how perception of the environment is related to identity processes, to relationships between people in social groups, and to processes of social representation, we can begin to see where the concept of sustainability is taking hold, and how it is guiding decisions and action.

There are a number of levels at which planning affects social processes. These will be discussed in relation to identity processes at the individual and group levels. It is because environmental considerations impact upon identity that reactions to planning for sustainability are so strong. Changes in the physical surroundings of individuals demand changes in the way they perceive those surroundings. In the framework outlined above, such changes amount to a requirement to adapt the representations of the surroundings which are currently held. If the individual's current group memberships include representations that attach strongly to aspects of the environment which are threatened by the change, then either the representation must be recast, or the change resisted so that the representation continues to hold. In either case, the identity of the individual as someone who 'lives in this kind of place, with people who see the world in this kind of way' comes under threat. At the group level, by the same token, environmental change, or indeed change in the framework of prevailing ideas about the environment (for example, the emergence of the idea of sustainability) similarly demand response on the part of groups, and generate instability in the relationships between groups. Some of the implications of these processes for urban environmental planning and sustainability will now be discussed at the individual and group levels.

Identity and individual responses to environment

Attitudes to the environment are integral to identity structures at both the individual and the group level. The individual constructs a self concept out of experiences of both social and physical location (Breakwell 1992). In the same way that a challenge to the social groups to which one has affiliation constitutes a threat to identity and provokes recognisable patterns of response (Tajfel 1982), so changes imposed upon valued physical environments provoke responses which are more than mere 'opinions about places'. The vehemence of reactions to even quite small changes to residential environments should alert us to the reality that more is being defended than bricks and mortar when planners are accused of attacking 'us' in a community. The immediate surroundings in which we are located define something about the identity we hold; it could be argued, for example that the high level of stress associated with moving house may have a component that is associated with the need to reconstitute oneself in a new setting, not only of new neighbours, but also of a new physical environment – hence the ritual of the house-warming. If this is so, then the implications of implementing new urban forms – such as the 'spread city' – in order to promote sustainability – are that there will be strong and deeply rooted responses to the planning of such change. Such considerations range well beyond the merely local and individual in their implications, as is cogently argued by contributors to the recent book by Duncan and Ley (1993). Ley, for example, explores ways in which architects for co-operative housing projects in Vancouver identified their task as being essentially one of expressing the identity of tenants in the built forms they produced. In particular, they reported that the individual participants sought not only a degree of uniqueness within projects but also, very markedly, an expression of group identity too. The development of 'landscape' and of the built form of the urban environment expresses power relations within society, but also becomes imbued with meanings that are not necessarily commensurate with the intentions of those who hold power.

One of the threatening dimensions of the emerging concern over sustainability lies in its power to evoke both new affiliations and new conceptions of what constitutes 'appropriate design'. We shall need to draw upon social psychological understandings of the response to identity threat if the planning of new urban forms is not to be wrecked on the rocks of the social response to such threat. One of the most important of these responses is the way in which group resistance to change is mobilised if individuals who feel threatened are able to recognise common interests with others. Changes to the physical environment constitute changes to the context within which identities are formed, thus imposed changes are likely to be understood as direct threats to identity. Empirical support for such an interpretation was given by the response of an inner city area undergoing renovation (Matthews 1981). Residents of the area had campaigned for renewal in their area, but as soon as the change began to be implemented, the community's existing pattern of social group relationships was realigned, leading to conflict between the residents and the local authority planners responsible for implementing change and to the emergence of new groupings and new conflicts within the local community. Interview data confirmed that over the first six months of the change, descriptions of the area became more elaborate, and evaluative statements more prevalent within those descriptions. All these results are commensurate with what takes place when groups and individuals experience threat: stronger differentiation between groups occurs, and the representations which members adopt are renegotiated and redefined. This would account for the stronger evaluative element in perceptions of the area. Conversely, the high levels of self esteem reported amongst residents of self-build homes, and amongst self-build communities such as that at Lightmoor (see Ekins 1992) can be interpreted as evidence of the consolidation of identities, and the absence of threat, and thus would seem to have much to tell us about the processes by which environmental change can be implemented successfully.

In relation to sustainability and the planning of cities, a number of points can be made. The concept of sustainability is increasingly part of the rhetorical culture of the planning profession. In developing strategies for its achievement, planners are being enjoined to develop effective public consultation – notably under the impetus of Agenda 21. Planners are, then, being called both to be explicit about the need for change in the design and use of the urban environment and to encourage public discussion of the shapes of such change. The outcome of such proposals and discussions can be predicted to be stormy on account of the strong threats both to group interests and to individual identities thereby engendered.

The group level of responses to environmental needs

If planning for sustainability is likely to generate individual responses to threat, as suggested above, there is also the second level of social and psychological operation to be considered; that of the groups which form reference points for individual responses and which themselves operate to establish representations about what is entailed by sustainability. There are important group processes operating both 'externally' to the planning system (setting the context for planners' actions), and 'internally' to it (establishing the social norms and values to which individuals working within the system must respond). Externally, urban residents in neighbourhoods and communities set one important group context for the reception of planning proposals. Internally, the various traditions and practices of planning, and the emergence of new specialisms and departments create the intergroup context within which the definition of sustainability, and the practices that are to achieve it must be negotiated. Such considerations have implications for the education of urban planners, some of which are discussed in the final section of this chapter.

For urban residents, local responses to planning decisions mesh with attitudes acquired from and through involvement in various groups. Many of these groups will have particular expectations about the way members respond to and behave in the physical environment. In this respect, the recent surge in support for 'environmental' organisations of various kinds is of particular interest, although groupings which are not explicitly 'environmental' in their concerns are also highly significant for the patterns of response to measures targeted toward sustainability. One of the important influences affecting the level of support for environmental objectives lies in the extent of knowledge possessed by the individual. Lyons and Breakwell (1992) conducted research which showed that amongst young people, social class and age were both positively correlated with environmental concern. Older respondents within the 13–16 age group, and those of higher socioeconomic class showed the greatest degree of environmental concern. However, over and above these factors, those young people who were most knowledgeable about scientific questions also showed the greatest concern over environmental issues, particularly where such knowledge was associated with worries about the speed and effects of scientific change. If young people are developing environmental attitudes within the context of their scientific education, it can be anticipated that measures for sustainability will increasingly find responses developed in the same context. Such findings are helpful in guiding us as to the degree of *concern* over environmental issues, but they do not give us a complete picture of the response to environmental *needs* of a direct and tangible kind. The importance of examining how strategies are communicated so that they are presented in ways that capitalise upon public awareness of scientific and technological capabilities is apparent. But we shall also have to go further, and explore what environmental needs evoke the concerns which are induced through the scientific education system. Clearly, there is also a continuing need both to improve the receptiveness of young people to science education, and to enhance their receptivity to environmental concerns through the medium of other aspects of the school curriculum. The status of environmental education as a 'cross curricular theme' within the Government's National Curriculum for schools derives considerable support from such findings.

Internally to the planning system, the social group dimension of environmental attitudes also allows us to locate explanations for the immense variations in the approaches of urban planners as practitioners to the goal of sustainability. Traditions in planning education, traditions of specialisation within the profession, and traditions in the critique of planning all produce sets of attitudes and perceptions on the part of planners which work against any consensus about what sustainability might imply, to say nothing of the question of the means by which it might be achieved. At the same time, however, the developing tendency for British local authorities to designate departments for environmental planning indicates the possible emergence of a new 'received wisdom' about what is entailed. The processes by which individuals become attached to such departments, or are nominated as operating outside them, give rise to the possibility of new and distinct definitions of the environmental agenda, and contain the danger that such definitions will be difficult to reconcile unless cross-departmental groupings can be established. The processes of inter-group differentiation and distinctiveness will need to be mobilised across different groupings if authorities are not to find environmental considerations corralled into specific areas and departments. Establishing environmental officers in each department, as has happened for example in the London Borough of Sutton, is probably helpful in this direction, and for these reasons.

Beyond this, regional and national variations in the nature and level of economic and resource development also imply quite disparate perceptions of what is required of a planning system and of what might be implied by a sustainable city. The recognition that group affiliation produces commitment to particular beliefs and attitudes, and that these beliefs and attitudes may well be elaborated so as to exaggerate differences between one's own and other apparently similar groups should alert us to another source of potential for conflict over achieving sustainable urban development. Again, the need is to understand how, and along what lines, the elaboration of differences occurs, so that the functions of such differentiation, which include the maintenance of group cohesion, can be fulfilled. If these functions are understood rather than deplored, it should become possible to mobilise them in the cause of achieving the superordinate goal of sustainability – although there remains the serious problem of definition inherent in the concept of sustainability itself.

Towards a strategy for adjusting the approach to sustainability

In view of what has been argued about the significance of group processes for producing variations of perception, attitude and action, the task that faces us may perhaps be more usefully cast as one which involves identifying and operationalising the particular definitions that are being used by the various participants in the pursuit of sustainability. If we can understand how different groups of practitioners prioritise aspects of environmental need, it may become possible to forge links between groups that have not previously recognised common concerns or complementary expertise.

At the level of group membership, therefore, we can begin to see how attention to the processes of group identification and cohesion help in understanding the strength of individuals' response to environmental issues. Such a perspective also offers a framework within which the effects of public knowledge and educational strategies can be interpreted. This focus also makes it possible to see how planners as practitioners will operate in ways that can be explained by reference to their professional group allegiances, and might be able to generate new patterns of interaction with other professional and lay groups. Evidence about the effectiveness of different information giving and persuasive strategies, in relation to the response of both individuals and groups, such as that discussed by Stevenson (1993), offers a strategy for action that responds to the group relationships identified in particular settings.

To move towards such an outcome, the approach of 'reflection in action' advocated for design practitioners by Donald Schon (1983, 1987) seems to have much to commend it. Schon's concern is with developing ways in which the education of professional practitioners can be made to include what he calls the 'reflection in action' that is embedded in skilful practice of a profession. He suggests that university training often gives privileged status to technical rationality and the application of privileged knowledge to instrumental problems of practice. His argument is that in reality, the skilled practitioner in whatever field works through a continuing interplay between technical knowledge and reflection on the particular and specific conditions of the problem being addressed, and he lays out an agenda for professional education which takes this as axiomatic, rather than suggesting that it is a kind of 'short cut' that comes to be used at a later stage in professional life, but which should not really be contemplated during training. In dealing with the problems of implementing urban planning decisions, such 'reflection in action' is an inevitable necessity, since every location, every client group, every community has its own peculiarities, and few decisions evoke precisely the response predicted for them. By recognising the patterns of group attitudes towards particular environmental concepts, and by being aware of the dynamics of inter- and intra-group responses to challenges to group attitudes, planners might heighten the degree of their 'reflection in action' and be better able to adjust the presentation or implementation of actions so as to capitalise on the energies of groups responding to change, or the need for change.

If planners are to become 'reflective practitioners' of the sort suggested here, there is an inevitable question about how this is to be achieved. We also need to ask whether reorientating planning toward a social analysis of its practice and away from an immediate pragmatic concern with developing specific local initiatives can reasonably be advocated in the light of immediate and pressing requirements for action. The current 'resurgence' of strategic planning (Breheny 1991) might also seem to call such a 'social' agenda into question. Part of the answer to such a challenge lies in the concept of reflection *in action*. What can be acknowledged is that the 'business' of planning for sustainability will be pursued by urban managers, planners, pressure groups and individuals to a greater or lesser extent depending on the particular circumstances of cities with different needs and of particular impacts of the economic climate. Policies for action, and for promoting environmental education are already in train and will continue. The debate about the inevitability of the link between environmental and economic priorities has yet to be joined in any pronounced way within 'popular consciousness', and the vast political agenda implied by the move towards sustainability and its inevitable redistributive effects have almost certainly not been recognised by the general public at large. But until a better appreciation of the pattern of such understandings is achieved, the potential for truly effective 'reflection in action' by planners remains seriously limited.

Research priorities for the profession, and for the academic analysts of the pursuit of sustainable cities, should lie in this area. The prescription is for an action research agenda in which the processes of social response within the urban management and planning system are identified and interpreted as the process of planning proceeds. The analysis of the impacts of decisions which is already necessary for the development of policies would be informed, in such an approach, not only by technical appraisal of the physical and institutional impacts of developments, but also by assessment of the effects of the decisions upon the attitudes and behaviour of those affected by planning both *in situ* and in the organisational structures of public and private agencies responsible for the implementation of action. The ultimate product of such research would be a better understanding of the impacts of particular plans, and perhaps better implementation of those plans. This approach offers the potential to develop a more effective theoretical framework for explaining the relationships between environmental planning, social responses to environmental need, and the practice of environmental management.

Sustainability objectives give prominence to long-term considerations. If such considerations appear to override immediate perspectives on housing needs, recreation or service provision, it will be very easy for the planners to be cast in the role of remote, privileged, and unknowing professionals, untouched by immediate daily concerns. The technologies which will assist in developing more sustainable urban systems are likely to add a further twist to any such perception, in that they will probably be seen as novel, 'technical' and remote from everyday experience. The task of winning support for technological innovation is thus another vital element in the search for sustainability strategies.

4

Sustainability: A Feminist Approach

MARY MELLOR

> The special message of ecofeminism is that when women suffer through both social domination and the domination of nature, most of life on this planet suffers and is threatened as well. (King 1989, p. 25.)

To say that women suffer as a result of the domination of nature is not to say that other groups do not suffer as well. In the United States African-American children are two to three times more likely to suffer from lead contamination than white children of the same income level, and African-American communities are three times more likely than white families to live near toxic waste dumps (Bullard 1993). Colonial exploitation and disruption of traditional forms of agriculture have destroyed the subsistence base of indigenous peoples in many parts of the world (Shiva 1989; Rau 1991). Workers, particularly in the unregulated export zones, suffer from high levels of industrial injury and illness (Mitter 1986). The argument made in this chapter could be made from the perspective of any of these groups; the exploitation of human by human and the exploitation of nature by human go hand in hand.

The importance of taking a feminist perspective rests in the argument that the fundamental division of power, and particularly of labour, between men and women, holds the key to the development of unsustainable patterns of development (Mellor 1992a;

Environmental Planning and Sustainability. Edited by S. Buckingham-Hatfield and B. Evans.
© 1996 by John Wiley & Sons Ltd.

Mies and Shiva 1993; Plumwood 1993; Seager 1993; Harcourt 1994). Women also suffer disproportionately where there are patterns of exploitation based on colonialism, racism or worker exploitation (Mies *et al.* 1988). The basic argument of this chapter will be that an understanding of the material position of women, in both industrial and non-industrial societies, will help us gain insights into the way in which those societies have become unsustainable ecologically.

To argue that women's experience is connected with, or in some way represents, that of the natural world is not without its dangers for feminists. Much of feminist struggle since the emergence of industrial society has been to dissociate women from nature and their relegation to home and hearth. Women have demanded, and to a limited extent achieved, their right to participate equally in the public world of politics, professions and commerce. However, while some women gained their place in the sun, many other women have slipped into ever increasing poverty. Women predominate among those in low-paid insecure work and women-headed families are among the poorest in western societies.

What has also become clear is that the promise of eventual equality through economic development and growth for not only women, but working class and racially and colonially oppressed peoples is not ecologically possible. It has, in fact, been an illusion (Douthwaite 1992; Mellor 1993).

The emergence of ecofeminism

> The brutalization and oppression of women is connected with the hatred of nature and with other forms of domination, and with ecological catastrophe. It is significant that feminism and ecology as social movements have emerged now as nature's revolt against domination plays itself out in human history and non-human nature at the same time. (King 1983, pp. 124–125.)

Ecofeminism emerged in the mid 1970s and brought together, as its name implies, feminism (primarily radical feminism, but also a number of socialist feminists) and ecology (leaning towards a dark rather than a light green perspective) (Plant 1989; Diamond and Orenstein 1990). Ecofeminists, like most greens, trace the destruction of the natural world to the hierarchical dualisms of western society. They see western socio-economic systems as prioritising culture as against nature; seeing the life of the mind as distinct from, and superior to, the life of the body; holding scientific knowledge as more valuable than traditional forms of 'folk' knowledge; claiming that abstract reason is separable from, and superior to, feelings and emotions; constructing a public world of politics, professionalism, commerce, production and social life that is separate from the private world of family and domestic relations, a division that gives supremacy to men and subordinates women (Plumwood 1993).

They argue that these dualisms have resulted in a one-dimensional public world devised by men in their image, embodying culture, mind, science and reason, eclipsing and suppressing women into a private world associated negatively with nature, body, folk knowledge and emotion. This division has allowed the unrestrained development of science and technology, industry and militarism. Male-dominated societies also exhibit destructive attitudes: aggression, competitiveness, single-path reasoning (following only one line of thinking as in military logic, economic calculation, or the scientific method). It has created huge unfeeling bureaucracies, atomised self-seeking and self-interested individuals and the amoral market-place. In turn it has eclipsed women's values of nurturing, caring and the need to sustain future generations.

Two early ecofeminist books traced the dominance of men over women and nature: Susan Griffin's *Women and Nature* in 1978 and Carolyn Merchant's *The Death of Nature* in 1980. The eclectic nature of ecofeminism is indicated by the fact that Susan Griffin has been identified with radical, cultural feminism, while Carolyn Merchant is a socialist ecofeminist. Griffin presents us with a poetic expression of the dualisms of western society, while Merchant gives us an historical exposition of how they occurred. Merchant lays the blame for the exploitation of women and nature at the feet of male-dominated religions and male-dominated science. She shows how the scientific revolution of the sixteenth and seventeenth centuries overthrew the previous organic view of nature that restrained exploitation. This led to the 'death of nature' in that it changed from being seen as a living organism, 'Mother Earth', to a lifeless machine which could be manipulated and exploited at will. Before the scientific revolution, the 'female principle' was uppermost in the interaction between humanity and nature which 'emphasized the interdependence among the parts of the human body, subordination of the individual to communal purposes in the family, community, and state, and vital life permeating the cosmos to the lowliest stone' (Merchant 1980, p. 1).

According to Merchant, the female principle was overthrown when the Judaeo-Christian God-given right to the domination of nature was embodied in the development of modern science. One of the main exponents of the new methods was Francis Bacon, who, as Merchant notes, was Attorney General to James I during a wave of witchcraft trials. Bacon's language is rife with the imagery of torture and persecution:

> For you have but to follow and, as it were, hound nature in her wanderings, and you will be able when you like to lead and drive her afterward to the same place again . . . . Neither ought a man to make scruple of entering and penetrating into these holes and corners, when the inquisition of truth is his whole object. (Quoted in Merchant 1980, p. 168.)

The earth was no longer alive and sacred, she was a passive object to be raped and pillaged. Mechanistic science saw the natural world as a series of machines in motion. The life history of natural structures was less important than the immediate benefits they could offer to 'Man' and the unchanging 'laws' of nature they could reveal. Irene Diamond and Gloria Orenstein argue that ecofeminism is recovering and revaluing the forms of knowledge that rationalistic science destroyed; that it is 'a new term for an ancient wisdom' (Diamond and Orenstein 1990, p. xv). They see the task for ecofeminism as being to envision another world, a whole world that 'heals' the one-sided destructive world that male domination has created. The beginnings of the search lie in women's knowledge and experience that men have repressed.

Such arguments must necessarily lead to a rethinking of the core tenets of both first-wave (nineteenth and early twentieth century) feminism and second-wave feminism (late twentieth century). Both saw women's identification with nature in male-dominated society as the basis of women's subordination and argued that the separation of women from nature was central to women's liberation. It was Simone de Beauvoir who, in 1949, first pointed to the way in which men treated both women and nature as the Other, to be both dominated and feared: 'Man seeks in woman the Other as Nature and as his fellow being. But we know what ambivalent feelings Nature inspires in man. He exploits her but she crushes him, he is born of her and he dies in her; she is the source of his being and the realm that he subjugates to his will' (De Beauvoir 1968, p. 144).

De Beauvoir saw this as a trap for women, which they must seek to escape. Women must be freed from nature. In her own life de Beauvoir refused to marry or have children and was proud of the fact that she had never learned to cook. Mariarosa Dalla Costa (1988) has argued that similar decisions are now increasingly being made by Italian women but for less politically committed reasons. Dalla Costa argues that the logic of living in a society that does not value women, their lives or their contribution to the future of human society, is that women refuse to bear children. In 1994 the Italian birth-rate was the lowest in the world at 1.3%.

For many greens concerned with sustainability this would seem to be a happy outcome, particularly those who stress the centrality of population control (Irvine and Ponton 1988). However, low population growth will make little difference to ecological sustainability if the remaining population of Italy and other industrialised countries continue to consume at their present rates. It would be several generations before a collapse in population would be sufficient to allow for a sustainable society at current levels of consumption.

Women in the industrialised countries making a personal decision not to have children are also in a very different position from women in the so-called developing countries who are subject to aggressive population control measures (Sen 1994). Population policy is clearly racist when the fertility rate of poor, low-consuming women of the South is deemed to be more crucial to the sustainability of the planet than the consumption patterns of the North. It is important to remember that Bangladesh, one of the most densely populated countries of the South, has a lower population density than Belgium (Moss 1994). Consumption patterns are critical to this debate. If the 5.5 billion people on this planet were all to live at the same standard as the average Californian, then we would need at least a couple of spare planets.

In practice, very few of the world's population live at that standard and only around one-fifth of the world's population have benefited substantially from industrialisation and 'development'. Even within rich countries like Britain and America, there are around a third of the population who cannot join in the bonanza because they fall below the poverty line (Galbraith 1992). Many of these are women. For ecofeminists, issues like population control are a diversion from the real issue of the nature of male-dominated socio-economic systems. It is the inequality and economic insecurity of women that leads to both the high population growth in the South and the collapse in the birth rate in the North. In bringing together ecology and feminism, ecofeminists see both women and nature, in the North and the South, as subject to the destructive technologies of modern patriarchal society: war and militarism, capitalism and industrialism, genetic and reproductive engineering. Most fundamentally, ecofeminists see the primary role of patriarchy in creating the modern scientific, industrial and military systems that are threatening the planet. They argue that men's attempt to 'master' nature is in danger of losing us the earth.

The tremendous changes that have taken place worldwide in the period since the emergence of industrialism have benefited only a minority of the human race, mainly men, mainly white. The values that have been favoured have been those of possessive individualism, the search for property and profit (MacPherson 1962). State-sponsored industrialism in Eastern Europe also prioritised male values and needs, as the lives of Soviet women showed (Hansson and Liden 1984). Ecofeminists argue that what we have seen is the emergence of an unsustainable destructively one-sided form of human development that prioritises an earth-destroying male culture at the expense of the nurturing and caring values associated with women. Any model of sustainability would therefore need to begin from women's lives and experience.

Irene Diamond and Gloria Orenstein (1990) see the emerging ecofeminist movement as a major catalyst of ethical, political, social and creative change that will 'reweave' the world. However, within ecofeminism there is a tension between those who see the relationship between women and nature as socially created, and those who see it as a deeper relation of biological and spiritual affinity that transcends particular societies and eras. Those who stress a 'natural' affinity tend to focus upon women's bodies, 'her womb, her roots, her natural rhythms' (Kelly 1984, p. 104) and/or woman's role as a mother, 'the experience of bringing forth and nourishing life' (Collard 1988, p. 102).

However, as I have argued elsewhere, this division is a false one (Mellor 1992b). Women are not closer to nature because of some elemental physiological or spiritual affinity but because of the social circumstances in which they find themselves, i.e. their material conditions. At the same time, women do have particular bodies which do particular things, but what matters is how society takes account of sexual differences and the whole question of the materiality of human existence. We not only have to be born, we have to find daily physical and emotional sustenance. We face both morbidity and mortality, we become sick, we grow old and die. What this amounts to is that we are *embodied*.

Whatever social lives we construct, they are always delimited and defined by our bodily existence. Equally we are delimited and defined by our environment, i.e. human society is *embedded* in its ecosystem. The basic argument of ecofeminism is that women, rather than men, have had to differentially bear the burden of the embodiedness and embeddedness of human society. They share this burden with racially oppressed, economically exploited and colonised peoples and the planet itself. What it boils down to is that a minority of the human race are able to live as if they are not embodied or embedded, as if they have no limits, because those limits are borne by others.

Whether they make their case on the basis of women's social exploitation and oppression or their affinity with nature, ecofeminists share a core of common ideas: the centrality of violence against both women and nature in the formation of industrial society and global economic systems, the sacrifice of women's health to the needs of production and the market system, the exploitation and commodification of both the natural world and women's bodies through genetic and reproductive technologies, the importance of the sexual division of labour as the material basis of male domination in both the North and the South, the importance of women's grass-roots struggles to preserve both their own lives and the survival of the planet.

Women in grass-roots struggles

One of the most interesting aspects of current campaigning around the issue of sustainability is the role of women and indigenous peoples in grass-roots struggles, from the Amazon Basin to the Himalayas, from New York to Kenya (Shiva 1989; Epstein 1993; Seager 1993; Braidotti *et al.* 1994). This is because for women and indigenous peoples the problem of sustainability has come 'close to home' (Shiva 1994). Two of the most well-known examples of grass-roots struggles are the campaigns over toxic waste dumps in the USA and the Chipko movement in the Himalayas.

Love Canal

Toxic waste has become a major issue in the United States where each year the country has to dispose of the equivalent of 2500 lb (1 lb = 0.454 kg) of hazardous waste for every man, woman and child (Newman 1994). The siting of the thousands of toxic waste dumps, most of them near working class, black and Hispanic communities, has become a major point of political campaigning. One of the most important outcomes of these campaigns is the politicisation and empowerment of local communities.

One of the earliest campaigners, Lois Gibb of Love Canal, New York, recalls that 'I grew up in a blue collar community, it was very patriotic . . . . I believed in government' (Krauss 1993, p. 111), that was until they refused to believe her when she claimed that her neighbourhood was built on a toxic waste dump. Lois Gibbs led a two-year struggle from 1978 to 1980, but it was not until women had vandalised a construction site, burned an effigy of the mayor and been arrested in a blockade that government officials began to take notice. Even then Lois Gibbs found that her evidence of the ill health of her own family and those around her was not taken seriously until she got a scientist to put her 'housewife data' into 'pi-squared and all that junk' (Seager 1993, p. 265). Women in other local campaigns also found themselves accused of being 'hysterical housewives' when they tried to raise issues about the dumping of waste. As one black woman from the American South put it, 'You're exactly right, I am hysterical. When it comes to matters of life and death, especially mine and my family's, I get hysterical' (Newman 1994, p. 58). There is now a national network in the USA to support local communities who find themselves threatened by toxic waste: the Citizens Clearinghouse for Hazardous Waste (CCHW), set up in 1981.

The Women's Environmental Network in Britain has also looked at the problems of toxicity, particularly dioxins which accumulate in the human body and can be passed on to children in breast milk (Costello *et al.* 1989).

The Chipko movement

> The forest is our mother's home, we will defend it with all our might (*Women of the village of Reni in the Garhwal mountains of the Himalayan Range*; Anand 1983, p. 182).
>
> What do the forests bear?
> Soil, water and pure air
> Soil, water and pure air
> Sustain the earth and all she bears
> (*Song of the Chipko women*; Shiva 1989, p. 77).

The Chipko movement of the Himalayas has been one of the most inspiring examples of women's affinity with nature, yet it is also an important example of the role of social and political action. The Chipko movement first came to public attention in 1974 when Himalayan women put their arms around their trees and hugged them to prevent them from being felled. In doing so they launched the Chipko (hugging) movement. They were not only expressing a spiritual unity with nature, they were demonstrating the way their daily lives depended on the trees for firewood and forage. While women in industrial communities may have to learn to develop an awareness of ecological issues, for poor rural women of the South this is the reality of their daily lives; all struggle is ecological struggle (Shiva 1989).

Despite its inspirational example, the Chipko movement was not simply a spontaneous flowering of women's immediate physical and spiritual identification with the forest. It grew out of the long and purposeful struggle of politically committed followers of Gandhi in the region (Shiva 1989, pp. 67–77). Inspired by Gandhi, one woman, Mira Behn, settled in the Himalayas in the 1940s and began to study the ecology of the region. Other women, like Sarala Behn and Bimala Behn, started ashrams for the education of hill women. The movement grew out of a 'mosaic of many events and multiple actors' (Shiva 1989, p. 68). It combined the traditional relationship of hill people to their environment, the political and spiritual commitment of followers of Gandhi, and the very immediate material needs of local women. Irene Dankelman and Joan Davidson (1988, p. 50) also tell us that in 1982–1983 the Dasohli Gram Swaraj Mandal organisation set up 20 eco-development camps in the region.

However, despite the wider political and ecological campaigning, women did play a special role and exhibit a clear affinity with the ecological needs of their region. Although the camps and the whole Chipko movement were open to both men and women, it was women who responded with the most appropriate understanding of the ecological issue and who undertook long-term committed action to defend the trees. The same pattern occurred in Africa where the Kenyan Green Belt movement started by Professor Wangari Maathai was organised through the National Council of Women. As in the Indian example, a central part of the programme was political and ecological education. Again, women responded readily with hundreds of local women's treeplanting groups being set up (Dankelman and Davidson 1988, p. 51).

What links women's grass-roots campaigns, North and South, is women's vulnerability to the problems of sustainability and their lack of access to the centres of decision-making which cause their problems. The importance of women's struggles in the North is that women bear the consequences of industrial production without being in a position to influence production decisions. Their response is always 'end-of-pipe'; they are not in a position to know or influence what goes into the pipe in the first place. When Greens argue that we must think globally but act locally they often overlook the fact that it is women who *live* locally (Mellor 1992c). Women have little choice but to think locally. They live near the waste dump, the poisoned well or the factory belching smoke. They are the people whose mobility is threatened by roads and traffic, whose children cannot play safely. They have to nurse the young, the old and the sick.

In the South, women are still fighting to retain their economic autonomy and access to their traditional sources of subsistence production. They are battling against loggers, commercial estates and dam-builders, to preserve their access to common land. They are struggling against poverty and forced sterilisation. For women of the North and South, in their isolation from centres of power and decision-making and in their experience of the consequences of decisions over which they have no control, they share with nature the experience of being an 'externality', that is something of which the economic system takes no notice and no account.

Outside the economy: women and nature as externalities

> The modern creation myth that male western minds propagate is based on the sacrifice of nature, women and the Third World. It is not merely the impoverishment of these excluded sectors that is the issue in the late twentieth century; it is the very dispensability of nature and non-industrial and non-commercial cultures that is at stake. Only the price on the market counts. (Shiva 1989, p. 221.)

> If human maintenance, mental and physical, and the nurturance of human beings are not taken care of, no other economy is possible. (Pietila 1987, p. 11.)

One of the most important boundaries that the 'economy' has constructed is between men and women's work. The 'market' is not just capitalist, it is male (Armstrong and Armstrong 1988). Most of women's work is unpaid because it is never bought or sold as a commodity on the market. Crops grown, meals prepared and clothes made are used directly, so that women do not produce a tangible 'product' other than a healthy child, a well-cooked meal, or a few gallons of water fetched from a distant well. On this basis most of women's work counts, quite literally, for nothing (Waring 1989). One of the most important achievements of feminist economics has been to gain recognition for the vast resource to societies all over the world created by women's domestic and subsistence labour (Lewenhak 1992).

As Marilyn Waring has forcibly argued, the invisibility of women, women's work and women's needs goes right to the heart of world decision-making systems, particularly the international calculation of comparative wealth, the United Nations System of National Accounts (UNSNA). Women's subsistence work (defined as 'primary work') had no value under UNSNA economic categories, 'since primary production and the consumption of their own produce by non-primary producers is of little or no importance' (Waring 1989, p. 78). Women's invisibility means that aid programmes often ignore women's needs for such fundamental things as access to clean water or sanitation (Dankelmen and Davison 1988; Rao 1989).

The strength of economic systems lies in the boundaries they are able to create and the boundaries they can destroy (Mellor 1992a). Economic systems carve themselves out of the whole complexity of social life. The capitalist market system in its present global triumphalism overruns the boundaries of self-sufficient traditional communities by destroying their economic autonomy. It is then able to scoop out the labour and resources of that community at grossly exploitative prices aided by racism and patriarchy. In the older industrial countries this process is already complete so that most people have no independent means of existence other than their 'freedom' to work. At the same time as it destroys the economic autonomy of peoples, 'the economy' retains its own autonomy. A boundary is placed around those things that are deemed 'economic', i.e. worthy of exchange in the market, or are acknowledged as the responsibility of economic systems.

What women do share with nature is their common exclusion from economic valuation. Women in their domestic and subsistence work and the planet's ability to withstand resource exploitation and pollution damage have been treated as externalities by formal economic systems. This is not just coincidence: the two forms of externality are directly connected as both are essential for human survival and the invisibility of women's work helps to obscure the damage of human activity to the planet.

What is women's work?

Women's work, underremunerated and undervalued as it is, is vital to the survival and ongoing reproduction of human beings in all societies. In food production and processing, in responsibility for fuel, water, health care, child-rearing, sanitation and the entire range of so-called basic needs, women's labour is dominant (Sen and Grown 1987, pp. 23–4).

Although women experience their relationship with men and with each other differently within and across different races, classes and cultures, there seem to be, in every human society, activities which are designated as 'women's work'. This does not mean that this work is done by *all* women in that society – the most privileged usually avoid it – but it is done by the majority of women and very rarely by men. While it is not popular in current poststructuralist thinking to talk of universals, there do seem to be some common elements in the type of tasks assigned to women. Throughout history and across cultures, women are overwhelmingly responsible for the basic needs, and particularly the personal care, of their families: children, siblings, husbands, older relatives.

The emergence of formal economic systems (state or market) has meant that a great many basic needs are increasingly being met in the formal sector. However, this does not mean that women's work does not exist. The male-orientated market/state economy which only recognises work if it is paid for, or makes a profit, disguises the existence and necessity of women's work. If it is not 'economic' to care for the elderly or provide a well for a rural village, because no profit can be created or the state cannot 'afford' it, the 'economy' washes its hands of the activity. This does not mean that the elderly are not cared for or that water is not fetched. In the last resort this work will be carried out by women. Women's work is necessary for the survival of both individuals and society. It is not work that 'needs' to be done to earn a wage; it is work that 'needs' to be done if human life is to carry on in any meaningful sense.

The essence of women's work is its 'immediate altruism' (Gilman 1915). It cannot be 'put off' or slotted into a work schedule. It cannot be 'logically' ordered or 'rationally programmed'. The needs that women respond to are unignorable demands; if they are ignored the social fabric of society begins to disintegrate. This work is altruistic in the sense that it is carried out for no reward other than pleasure in family relationships or through obligation, a sense of duty (Schreiner 1978). Throughout the world the majority of women are trapped into patterns of obligated labour through male-dominated family structures and male-dominated economic systems. Their altruism although spontaneous and often freely given is also imposed on women as a gender (Mellor 1992a, p. 252). There is nothing intrinsic about women's work that means a woman should do it, but the most important thing about this work is that somebody has to do it. Women's unpaid work is the altruistic work that keeps the whole society functioning; it is the work that creates 'humanity' both as individuals and as a community.

At the same time, women's primary responsibilty for this work is not inherent in their biology. Women give birth and nurture, but the biological capacity to give birth is so heavily influenced by the social context it cannot be argued that women are 'naturally' altruistic, loving and supportive. Biology is reality, but it is not destiny. Nor do all women do women's work or all men avoid nurturing and caring work. What is important is that male-dominated society has created a world which assumes that this is the case. The imposition of altruistic behaviour upon women is, I would argue, the most destructive division in human society. It means that the public world based upon male experience is quite literally a ME-world, constructed on the false premise of an independently functioning individual, with the nurturing, caring and supportive world hidden, unpaid and unacknowledged. Even when women carry their nurturing tasks into the public world, they are lower paid and have less status than men.

The division of labour is not just in what men and women do, but the differences in their income, life-styles and power. The most important separation in industrial development was between home and work, between the public world of waged work and public life and the private world of the family and household chores. In pre-industrial societies families worked close to home and shared common tasks, particularly in agricultural work. In consequence they were embedded in their environment, in both work and leisure. Life was, however, harder for women as they not only had to share in 'outdoor' work, they were also responsible for domestic work and childcare (Coontz and Henderson 1986; Miles 1988).

This remains the case for the majority of women in the world who are still involved in subsistence work. Not only are they responsible for the sustenance of their families, but Vandana Shiva argues, for the fertility and survival of their natural environment (1989). Shiva argues forcefully that women as the primary producers of subsistence have accumulated an invaluable store of knowledge that is in danger of being obliterated as the global market economy drives women from their traditional common lands and their subsistence ways of life. Women's knowledge as subsistence farmers embraces the diversity of biological and ecological knowledge: plant species, soil conditions, response to seasonal changes, water sources, pest control. Not only is this knowledge being lost but the diversity of the environment itself is being extinguished by industrial farming. The one-dimensional profit-orientated mentality of western science and technology is not only creating monocultures on the land, but monocultures of our minds (Shiva 1993a).

Vandana Shiva can, perhaps, be accused of over-romanticising the life of, and overstating the importance of, the woman subsistence farmer, but this does not affect the burden of her accusation against the global market economy and the dominance of western scientific thinking.

Conclusion

The ecofeminist case for linking the experience of women with ecological sustainability does not rest on women's essential and universal identity with 'nature' either as biology or ecology. Rather it rests on women's material reality and the pivotal position they play in mediating the relationship between male-dominated economic and social systems and the embodiedness and embeddedness of human societies. This experience differs as between the North and the South and within different societies depending on the social position and social marginalisation of different women, but women's subordinate position is universal enough to undermine the poststructuralist claims that no such category exists (Riley 1988).

Women are the first as well as the last 'colony' to be exploited by male-dominated economies (Mies *et al*. 1988). Even in the cultures of 'high modernity' women create the space for men to exist. As I have argued in *Breaking the Boundaries* (Mellor 1992a), 'social' time, i.e. socially available time for paid work, public life and leisure, has to be created out of biological time, the time it takes to sustain our physical and emotional existence. This is not only the daily routine of subsistence and cleanliness but our immaturity, morbidity and mortality. While it is not true that 'anatomy is destiny' in Freud's terms, it is reality. Women's identification with the 'natural' is not evidence of some timeless unchanging essence, but of the material exploitation of women's work, often without reward. It is not even so much the work that women do, but their availability. Someone has got to live in biological time, to be available for the crisis, the unexpected as well as the routine.

By placing on women the major responsibility for nurturing and caring values and activities, a public world has been created in the one-sided, distorted and damaging image of male experience. What has been created is a male experience world, a ME-world, that excludes women's experience and particularly the work that women do. As a consequence this world also obscures its own frailty, i.e. its embeddedness and embodiedness in its own biological and ecological existence. This, in turn, obscures its unsustainability. Ecofeminism exposes the unsustainability of western 'development' by showing the way it has been built upon the interests and experience of privileged men. It exposes the hidden reality of embodiment and embeddedness that male-dominated socio-economic systems deny. As such it speaks from the standpoint of the oppressed and exploited in such a way that the material reality of the situation is exposed (Hartsock 1984; Harding 1993).

Ecofeminism tells us is that it is not possible to transcend the materiality of the body. It may be possible for some people but only at the expense of those who carry the consequences. In that women bear the main responsibility for our natural being, they can also be seen as representing our connection with nature. This is not because women are somehow special or different from men in some essential sense, but they differ in the nature of the work that is associated with them. Equally where the burden of existence is carried by other groups their experience can also help us understand the underpinnings of human existence: the bonded labourer, the indigenous peoples thrown off their land, people who are marginalised through racialism or social class, who live near the toxic waste dumps and in the areas of highest pollution, who are exposed to hazardous processes in their work.

Sustainability can only be attained if we learn to live within the limits of our embodiment in our biological existence and our embeddedness within our environment. Because of the male-domination of western economic and scientific thinking, human existence and development was presented as unlimited and unconstrained. It promised that time and space would be created for all in due course as a result of progress, growth and development. The ecological crisis shows this promise to be a sham, as does the hidden life of human frailty that women's work represents. It also undermines the arrogant claims of western science and technology. 'Man' cannot create to the extent that he can destroy.

A sustainable society would need to incorporate the hidden work, interests and experience of women. It would be a women's experience world, a WE-world. A WE-world, based upon women's interests and experience, would need to be both decentralised and safe – ecologically, physically and socially. Domestic life and paid work would have to be integrated as women cannot move far from their homes because of their domestic and caring responsibilities. The boundaries between paid and unpaid work are in any event anachronistic from a woman's perspective. As all production would need to take place near the home, dangerous, polluting work could not be carried out as it would affect the health of the local community. People would also not want their local environment disfigured by destructive forms of production. Shops and other social facilities would need to be within easy reach of the local community and not stuck on the edge of town available only to the car driver. People and not traffic would have priority on the roads. Public transport would be universally available.

A world based on women's experience would ensure people's physical and social safety. Personal violence would never be treated as a 'private' matter. Streets and public spaces would be patrolled if there was any danger to children, minority groups or women. A WE-world would see its primary role as producing well-rounded and creative human beings and a well-integrated community life. People would not be asked to uproot themselves every few years to chase promotion or to meet the needs of a transnational corporation. If someone were needed in a new location a local person would have to be appointed and trained. People would not be forced to migrate hundreds or thousands of miles to look for work, never seeing their family or community for years on end. Emotional needs would be given equal priority with physical needs; people would listen to each other, sympathise and empathise.

This is not to romanticise women. Not all women are sympathetic and empathetic. Not all women want to live sustainable lives. I am not making claims for the inherently superior nature of women as the basis of sustainability. What I am saying is that male-dominated western societies have created a false image of themselves and their sustainability and that the sexual division of labour and the subordination of women has enabled this to happen.

5

Black on Green: Race, Ethnicity and the Environment

JULIAN AGYEMAN and BOB EVANS

A student of contemporary environmental politics and issues, faced with a representative range of current books and articles on the subject, might be forgiven for thinking that 'the environment', and 'environmentalism' are colour blind – that questions of race and ethnicity are of peripheral importance within the global, national and local conflict over our common environmental destiny. To be sure, the environmental literature is increasingly concerned to address questions of social justice, and many of the recent contributions to the environmental debate (e.g. Pepper 1993) quite rightly place equity and the unequal distribution of political and economic power centre stage. Even the official global plan for sustainable development, the enigmatically named 'Agenda 21', recognises the importance of questions of democracy and egalitarianism as fundamental to long-term environmental action.

However, although Agenda 21 makes reference to 'indigenous peoples', this document, like most current writing on the environment, fails to recognise the particular significance of race and ethnicity as important components of the environmental debate. The aim of this chapter, therefore, is to seek to redress this position, and our central premise will be that the environment is *not* colour blind. On the contrary, we wish to argue that questions of race and ethnicity are indivisible from environmental questions, not least

Environmental Planning and Sustainability. Edited by S. Buckingham-Hatfield and B. Evans.
© 1996 by John Wiley & Sons Ltd.

because environmental exploitation is invariably linked to economic exploitation, and racism is an increasingly important part of this global process.

We recognise that the concepts of 'race', 'racism' and 'ethnicity' are not unproblematic, and that there is a vast and contested literature which seeks to deal with them, which we are unable to review here. However, we do need to emphasise that we broadly agree with the perspective articulated by Miles (1993). We take it as axiomatic that 'race' is a socially determined rather than a biological category, and that the 'race relations' paradigm which dominated analysis of these issues in the 1960s and 1970s is an ideological construct which assumes the existence of discrete 'races' whose 'relations' require manipulation and management to secure stability.

In contrast, Miles and others (e.g. Sivanandan 1988) emphasise that racism is but one mechanism which justifies and legitimates social and economic exclusion, and that, furthermore, it is not only black people who are the victims of racism. Miles argues that, in addition to colonial racisms, other racisms, internal to Europe, exclude and subordinate other Europeans. For Miles, racism rather than race or ethnicity is the central analytical concept: 'The nation states of the EC are not confronted with a "race problem", but rather with the problem of racism, a problem which requires us to map and explain a particular instance of exclusion, simultaneously in its specificity and in its articulation with a multiplicity of other forms of exclusion' (Miles 1993, p. 23).

In the context of our particular task, which is to examine questions of race and ethnicity and the environment, we therefore take as our starting point the view that subordination, and exclusion by dominant groups of other groups on the basis of skin colour, religion, culture, ethnicity or any alternative conception of 'otherness', is a part of a process of racism which has a clear environmental dimension. Whether it is the 'environmental racism' of the dumping of toxic waste in 'third world' countries by the prosperous North; the massive open-cast coal mining in Colombia with its attendant child labour which provides cheap fuel for northern power stations; or the effective exclusion of black people from the British countryside – all these are the environmental manifestations of social and economic exclusion and subordination based upon racism.

In the remainder of this chapter we focus upon three elements of this process, whilst at the same time recognising that there are many more examples and cases. First, we examine the global context, arguing that 'environmental racism' must be understood as more than specific incidents of shipping hazardous waste, or polluting of particular communities. Following the dictum to 'think globally, act locally', we then turn to local environmental conditions where for many, if not most, black people, their local environment is poor relative to the world's white population. Finally, we assess the phenomenon of black environmentalism, both in terms of the differing interpretations which different social groups may place upon the environment, and the ways in which black and ethnic minority groups are addressing environmental exploitation.

A global context

The 1992 'Earth Summit' was, in many ways, a major watershed in the development of environmental policy. Although the outcomes of the summit were very limited (non-existent some claimed) in terms of concrete policy and action, the Rio meeting formalised what environmentalists had been saying for years – that both global thinking and local action will be necessary to deal with the emergent environmental problems of the earth. However, despite the alleged recognition of this principle, the Rio Summit also revealed, even more clearly, the reasons why many of these problems will remain unsolved for the foreseeable future.

The rhetoric of Rio is compelling. The environmental problems facing humankind are of such a magnitude, and such interconnected complexity, that they can only be solved through international co-operation. The sharing of common futures and fates is a prerequisite for the achievement of global sustainability, and this must be worked for through local action. Increased democratisation and citizen participation; the development of local capabilities through 'capacity building'; the active involvement of hitherto disadvantaged groups (women, young people and indigenous peoples); the development of strategies to 'empower' local people to take the responsibility for their own environmental futures – these notions feature time and time again throughout the 700-page document which is Agenda 21, the agreed global strategy for sustainable development.

The reality of Rio was, of course, somewhat different. The conflict between the generally prosperous North and the relatively poor South inevitably dominated the proceedings, as it was to do again in the subsequent United Nations conferences on population control (Cairo in 1994), climate change (Berlin in 1995) and social development (Copenhagen in 1995). The North is, in general, happy to take environmental action that does not significantly damage markets, sources of raw materials, productivity or profitability. It is also happy to press for environmental action in Southern countries which are deemed to have a poor environmental record. In contrast, of course, the South sees no reason to adopt punitive environmental measures when, after all, it is the North that is causing most of the world's environmental damage, whether this is defined in terms of contributions to greenhouse gas production, the creation of toxic waste, or the rapid utilisation of non-renewable energy sources.

All this is unsurprising and well-documented, but it has to be recognised that there is a clear racial dimension here. Standards that are deemed to be inappropriate, or even dangerous, in Northern, largely white societies, are often considered by representatives of those same societies to be perfectly acceptable in the mainly black South. The leakage of poisonous gas (methyl isocyanate and 23 other chemicals) from a tank at the Union Carbide pesticide factory near Bhopal, central India, in 1984 resulted in an estimated 6600 deaths and up to 600 000 injured. The precise cause of the incident is unclear, but it seems likely that the owners of the factory, the American multinational Union Carbide Corporation, would first have been unable to easily site such a facility in a Northern country, and secondly would have been forced to pay massive compensation to the victims of the incident. In India, the average payout is £500 and many thousands have yet to be compensated (Urquhart 1994).

The well-entrenched belief endemic to Northern societies is that the extant economic and social conditions of many Southern societies legitimately permit commercial behaviour that would not be accepted in the North, often justified in terms of bringing employment to areas of no work. However, the implicit, unstated position is fundamentally racist, in that such commercial behaviour is deemed as appropriate for black people whereas it is inappropriate for white people. It is far away from the economic power centres of the North, and any local representations concerning pollution or environmental degradation are likely to be muted for fear of unemployment.

The shipping of toxic waste from the North to West Africa or Mexico is more than simply an economic transaction and the French decision to resume nuclear testing in the Pacific is not simply a scientific matter. Both cases involve the suspension of ethical standards since these are deemed not to apply to 'others,' i.e. non-whites, non-Europeans. Racism is one implicit justification for such environmental and human exploitation. In this sense, 'environmental racism' must be understood as more than a simple process of environmental dumping in 'third world' countries, or the locating of exploitative and polluting industrial plant in black, Hispanic or native American communities in the US, the South or the Ukraine. Environmental racism has a much wider purview in that it is best understood as a perspective which both creates and legitimises environmental degradation and poor living and working environments globally for those 'others' who are viewed as economically, politically and socially subordinate. In this sense, environmental racism is as much about urban decay and neglect in Handsworth, Birmingham, as it is about the export of industrial waste to Guinea-Bissau.

Local environmental conditions

It is clear that the majority of the world's black population experience a local environment, within which they live and work, which is inferior, often dramatically so, to the one enjoyed by most white people, whether that environment is in Kingston, Jamaica, in Soweto, South Africa, or in Brixton, UK. The global process of capital accumulation has privileged the mainly Northern, mainly white, metropolitan countries, over the mainly Southern, mainly black periphery. Moreover, the international migrations of labour, particularly those which have occurred during the last century, have enabled 'third world' conditions to be established in the heart of the metropolitan North, creating a 'South in the North'.

In the case of Britain (but it could equally well be France, Germany, Canada or the United States) black people experience poorer housing conditions, have more restricted job opportunities and higher rates of unemployment, receive lower household incomes, tend to fare less well in education, and, on virtually any other indicator chosen, tend to experience poorer conditions and circumstances than the white population. This situation is well documented (e.g. Skellington *et al.* 1992) and does not require repetition here. However, what does require emphasis is the consequence of this in terms of local environmental quality.

Britain's black population is largely concentrated in urban areas, and within these areas it is characterised by residential segregation into the older, inner city areas (Peach 1986; Smith 1989). Approximately 75% of Britain's black and Asian residents live in a set of enumeration districts which contain only 10% of whites. In contrast, only 3% of black and Asian residents live in British rural enumeration districts, compared with a national average of 24% (Skellington *et al.* 1992, p. 44). This distribution has permitted the perpetuation of a popular mythology of 'black slums', or even 'ghettos', fuelled by the popular press at times of unrest or disturbance such as happened in Brixton, London, and St Pauls, Bristol, in the early 1980s, or more recently on the Broadwater Farm Estate in Tottenham, London, some ten years later. As Ward (1989) has pointed out in the North American context, the ethnically identifiable poor came to define the 'inner city' (Cross and Keith 1993).

However, whilst it would not be accurate to designate all inner city areas as experiencing poor environmental quality, it is broadly the case that those British inner city areas that are home to low-income groups generally experience poorer environmental conditions, and furthermore, this is particularly true of those areas which have a high proportion of black and ethnic minority residents. Apart from poor housing conditions, such areas are likely to experience higher levels of pollution from motor vehicles, lower levels of provision of public open space and 'greenery', poorer social facilities including the quality of educational and other public buildings, higher levels of litter and street detritus, and so on (Agyeman 1989). These are the environmental, as opposed to the social factors, which may encourage those with the option of mobility, both black and white, to move elsewhere in the search for a 'better area'.

All these factors could be interpreted as a simple consequence of the generally low social and economic position of Britain's black and ethnic minority community – that as new migrants from the Caribbean and the Asian subcontinent to Britain in the 1950s and 1960s took the jobs that whites did not want, they also took the housing that was no longer attractive to indigenous workers. However, it is clear that the maintenance of social, political and economic subordination of migrant workers and their descendants is in major part a consequence of social and institutional racism.

One element of this process is the operation of the public planning system for land use and other environmental activities. As Goldberg (1993) notes, urban planning has a long history of racism, quoting the way in which European colonists established 'native' and European quarters of colonial cities initially in order to protect whites from the plague, but more significantly as a mechanism to prevent the 'contamination' of Europeans by 'uncivilised' natives. Goldberg goes on to argue that the active state intervention in the post-war urban development of Western cities, as of colonial cities, was encouraged by means of nuisance law and zoning policy, to guarantee the most efficient ordering and use of resources: 'The principle of racialised urban segregation accordingly insinuated itself into the definition of post-colonial city space in the West, just as it continued to inform post-independence urban planning in Africa' (Goldberg 1993, p. 49).

The land use planning system in the UK has tended specifically to benefit those who own land and property, in addition to providing more general benefits to the articulate and educated middle class (Reade 1987). In this situation, it is not surprising to find that black and ethnic minorities have faired poorly in comparison with the aforementioned groups. A report on race and land use planning published in 1983 (CRE/RTPI 1983) detailed the difficulties faced by Britain's black communities when dealing with the land use planning system, and suggested some ways in which these might be overcome. However, perhaps not surprisingly, a follow-up study undertaken ten years later (Krishnarayan and Thomas 1993) confirmed that little has changed.

The environmental experience of black and ethnic minority groups in the United States is better documented, but the same circumstances apply. Poor environmental quality is almost always associated with those geographical locations with high percentages of African-American residents. Moreover, as Cutter (1995) reports, race is a better predictor of proximity to commercial hazardous waste facilities than low income. Minority residents in Detroit, for example, are four times more likely to live within a mile of a hazardous waste facility than white residents (Cutter 1995, p. 116).

This situation has prompted the rapid growth of the 'environmental justice movement' in the United States (see below), directed towards securing greater environmental equity, and this has been a major factor in the recent decision by President Clinton to sign an Executive Order requiring every federal agency to achieve the principle of environmental justice by addressing and ameliorating the human health or environmental effects of the agency's programmes and policies on minority and low-income populations in the US (Cutter 1995). As with all such declarations of intent, there is likely to be a gap between the rhetoric and the reality of implementation, but the policy nevertheless represents a major shift in terms of official recognition of the process of environmental racism. In the next section we examine the development of the 'black environmentalism' which has begun to effect this shift of attitude, and to assess the different interpretations of 'the environment' which has hitherto been dominated by white, and as Mellor correctly states, male perspectives (Mellor 1992a).

Black environmentalism

Is there a differentiated 'black environmentalism', or is it a reaction to the aims and objectives of environmentalism generally, which, as we explained above, is not 'colour blind'? On the one hand, it is tempting to say that there is a differentiated black environmentalism, with the formation of groups such as the Black Environment Network (BEN) in the UK (Agyeman *et al.* 1991) who, from their inception in 1988, began to introduce the concepts of equality and social justice into environmental discourses, and the more organised 'environmental justice movement', a protest coalition linked to the Civil Rights movement in the US.

On the other hand, the relative infancy of 'black environmentalism' in an organised fashion, especially in the UK, has made it predominantly reactive to the mainstream, but not entirely so. As Taylor (1993, p. 287) notes, in her study of minority environmental activism in Britain, 'their (BEN's) agenda is now driven by what pleases other mainstream environmental organisations and what will get funded'. Whilst this was not always the case (see Agyeman (1989) for issues related to countryside access), BEN's agenda lately does seem to be more driven by funding, rather than following a clear, community-driven alternative agenda.

Are there any characteristics of black environmentalism? Again, this is a difficult question, given that most of the (US) research has been carried out by whites, based on issues which they define as 'environmentalism'. For instance, there is evidence from the US that different ethnic groups interpret and value environments in different ways (Van Ardsol et al. 1965; Hohn 1976; Kellert and Westerfelt 1983; Kellert 1984; Kaplan and Talbot 1988; Baas et al. 1993), and from the UK (Agyeman 1989; Malik 1992; Environ 1994). However, again, this area of research is less well developed in the UK. Indeed, as Agyeman (1989, p. 337) notes, 'how much . . . [research] has been done by the Department of the Environment on the environmental quality enjoyed [sic] by black people? A Department official, when asked this question, replied "it's not the kind of research we do. It doesn't come under our auspices." '

Baas et al. (1993, p. 526) have argued that 'ethnicity may play a role in the type and level of expectations people have towards natural recreation sites'. This is similar to Kaplan and Talbot's (1988) findings, where preference varied by racial group, with African-Americans preferring more built-up environments. In the UK, Environ (1994, p. i) found that amongst Asians in Leicester, 'crime, unemployment, the economy and racism are of the greatest concern, but the environment is firmly on the agenda, particularly the quality of local living space (clean, uncongested streets and pollution free air and water). Asian people are more likely to become involved in initiatives relating to these issues'. It would seem that, in both the USA and UK, local or 'doorstep' environmental issues, especially pollution-related issues, are of the greatest concern amongst ethnic minority groups. But are these groups active in such issues, as part of the mainstream environmental movement?

Historically, both 'the environment' as a concept, and the environmental movement as a construct, as we have argued above, have been seen as 'colour blind'. Moreover, UK research on participation in the environmental movement (e.g. Cotgrove and Duff 1980; Lowe and Goyder 1983; Micklewright 1987; Agyeman 1988) has concentrated on class, with Lowe and Goyder (1983, p. 10) noting that 'the membership of environmental groups is predominantly middle class'. They follow this statement with an important distinction between *membership* (i.e. *joining* a group), and *interest* in the issues, which they argue 'is widely spread across all social classes' (Lowe and Goyder 1983, p. 13). If class does affect the likelihood of a person *joining* an environmental group, but not necessarily their level of *interest* in the issues, how does race affect *interest* and *participation*?

There is as yet little UK empirical research *specifically* into race and the environment. Building on the work of Agyeman (1989), Malik (1992) investigated working class inner city white and ethnic minority attitudes towards the countryside, and compared these to suburban middle class ethnic minority and white attitudes. She found that whilst both the white and ethnic minority inner city residents were less likely to use the countryside than their white and ethnic minority suburban counterparts (i.e. class was important), there were significant racial differences in terms of obtaining information (word of mouth was very significant for inner city ethnic minorities) and choosing where to go (Snowdonia and the Lake District were popular with suburban Indians because it reminded them of Indian hill forts). Agyeman (1988, p. 3) has called the environmental participation differences between working class white, and ethnic minorities 'working class plus' differences: the plus (i.e. the use of word of mouth information transfer) in this case being racial, and linked to language barriers.

Bullard (1993b) and Taylor (1989) have comprehensively assessed ethnic minority (especially black) participation in the USA. Taylor (1989) looked at what she calls the *'concern gap'* between blacks and whites through the relationship between environmental *concern* and environmental *action*. She notes that:

a lower percentage of blacks considered themselves sympathetic toward and active in the environmental movement than whites (Mitchell 1980), and that blacks/non-whites were less likely to perceive environmental hazards such as smog (Hohn 1976) than whites even when they lived in severely polluted areas (Van Ardsol, Sabagh and Alexander 1965). Kellert (1984) found black adults to be substantially less interested, concerned, and informed about the natural environment than whites, and Kellert and Westerfelt (1983) found non-white children to be less knowledgeable about, and less interested in, wildlife. (Taylor 1989, p. 176.)

Taylor (1989) then goes on to look at the available explanations for this environmental 'concern' gap between blacks and whites by reference to three categories of study: social-psychological (e.g. socio-economic status, marginality and hierarchy of needs), cultural (e.g. mythology, slavery and ethnicity) and measurement error (e.g. inappropriate indicators and sampling). She notes that 'these studies . . . indicate that the level of concern of blacks for the environment is consistently lower than that of whites. This "concern gap" is paralleled by an "action gap", that is, concern does not necessarily translate directly into action (Mohai 1985), therefore concern may result in some action or no action at all' (Taylor 1989, p. 180).

The way forward (i.e. in order to increase black participation), argues Taylor (1989, p. 199), is to meet the following prerequisites:

1. Solidarity
2. Cognitive perception of reality
   (a) develop political efficacy
   (b) recognise advocacy channels
3. Resources
   (a) monetary
   (b) knowledge (political expertise)
   (c) moral
4. Psychological factors
   (a) ideology
   (b) discontent

Solidarity, she argues, comes from black people's high levels of affiliation with social, political, and religious organisations. Low levels of political efficacy, and non-recognition of advocacy channels, she argues, are problematic. In terms of resources, again, there are problems, especially in terms of fiscal resources and a political knowledge base. In psychological terms, Taylor (1989, p. 200) argues that 'the environmental movement could attract more blacks through an expansion of the civil rights agenda to include environmental issues'. Just how Taylor expects the environmental movement to expand the civil rights agenda is not clear, and, whilst not directly applicable in Britain, it is clear that organisations in what could be called an embryonic civil rights movement, such as the National Black Caucus and Society of Black Lawyers, have never broadened their agendas to include environmental issues. Therefore, inverting Taylor's (1989) argument, we would argue that the embryonic British civil rights agenda could be expanded to develop a greater black environmentalism.

One area where ethnic minorities in the US are active in environmental campaigning is in relation to toxic waste dumping. Whilst the majority of western environmental campaigning on this issue has focused on dumping in the Third World, the US Commission for Racial Justice (1987) found that although socio-economic status was important in the location of commercial hazardous waste facilities, race was the key factor and that three of the five largest commercial hazardous waste landfills were located in black or Hispanic communities. Bullard notes that:

> activists of color have begun to challenge both the industrial polluters and the often indifferent mainstream environmental movement by actively fighting environmental threats in their communities and raising the call for environmental justice. This ground swell of environmental activism in African-American, Latino, Asian, Pacific Islander and Native American communities is emerging all across the country. While rarely listed in the standard environmental and conservation directories, grassroots environmental justice groups have sprung up from Maine to Louisiana and Alaska. These grassroots groups have organised themselves around waste-facility siting. (Bullard 1993, p. 24.)

Of increasing importance in the US, are the concepts of *environmental justice* and *environmental racism*. The latter term was developed by the Commission for Racial Justice (1987) to explain the disproportionate risk faced by ethnic minorities in their findings as regards the siting of toxic waste facilities. The former term is now used as a byword for black environmental activism in the USA.

At the time that the Commission for Racial Justice in the USA was developing its critique of the mainstream US environmental movement, which, it argued, ignored the siting of toxic waste facilities, a group in the UK, the Black Environment Network (BEN), was being formed. Articulating the need for environmental discourses to be informed by the concepts of equality and social justice (Agyeman 1989), BEN found common ground with US environmental justice groups. At this time, UK mainstream environmental organisations seemed to be unaware of the importance of these issues. However, as we noted above, BEN has now fallen into the trap of becoming 'accommodated' by the mainstream environmental movement. It has become funding led, and has lost the critical edge that it had in the 1980s.

The centrality of the concept of sustainability, especially after the Rio Summit in 1992, focused environmentalists' attention on notions of equity and social justice, areas in which they had shown little interest before. As Edwards (1992, p. 50) argues, 'it may surprise some people to know that a key reason why people of color may enthusiastically embrace sustainable development is because they hope that it is a new road that will lead to an old objective – a United States of America transformed by the guiding principles of freedom, justice and equality'. Not only did UNCED help to focus environmentalists' attention on these core concepts, but its major outcome, Agenda 21, called for greater participation of women, youth and indigenous peoples (i.e. under-represented groups) in the challenge of sustainability.

Through the local expression of Agenda 21, i.e. Local Agenda 21, local authorities are now developing, with their communities and businesses, consensual plans for local sustainability. Given this greater awareness of social justice and equity issues, are consultation mechanisms being adopted by local authorities which will reach under-represented groups? In a recent survey of Local Agenda 21 initiatives amongst UK local authorities, ERIC (1995) found that 66.2% of authorities had not adopted new public consultation procedures in order to fulfil the community participation requirements of Local Agenda 21, whereas 33.4% had. This would suggest that in the majority of UK local authorities, traditional consultation mechanisms will be pursued. As has been argued elsewhere, these will not reach under-represented groups (Agyeman and Evans 1995).

There are clearly initiatives in Britain, which, if built upon, might constitute a differentiated black environmentalism. This would require a re-focusing of organisations such as BEN and the building of greater links with the 'embryonic civil rights movement'. As Taylor (1993, p. 287) notes, 'they (BEN) need those links so that the ideas they are articulating can be expressed more forcefully'. In the US, these links are already made, and the result is a differentiated 'environmental justice movement' with its own alternative agenda for ethnic minority environmental action.

Race, sustainability and environmental planning

The current environmental policy debate is overwhelmingly influenced by the rhetoric of Rio. Agenda 21 speaks of participation, partnership, democratisation and community as central organising concepts in the search for sustainability, but as the previous pages of this chapter have demonstrated, for the majority of the world's population, the processes of racism and other forms of economic, social and political subordination will continue to ensure their effective exclusion from healthy, clean and unpolluted environments. On the one hand, the logic of the rhetoric of Rio is completely correct. Sustainability is only likely to be achievable if there is some equality of experience and outcome, and some sharing of common futures and fates. However, on the other hand, the Rio discourse of sustainable development is being conducted within the context of a global capitalist economy which requires and demands the subordination of nations and social groups as an inherent part of the process of accumulation. A recognition of this context does not necessarily have to result in paralysis or inaction, but it does require that those involved in the process of environmental planning recognise the complex realities of, in this case, deeply rooted racism and its environmental consequences.

A greater understanding and appreciation of these processes of racism and their consequences by environmental policy-makers – a recognition that the environment is not 'colour blind' – will undoubtedly assist in implementing the spirit of Rio. Moreover, an understanding that there is more than one interpretation of 'the environment', as BEN and others have shown, may encourage environmental planners to develop more comprehensive consultation strategies which will result in more representative and more widely supported policies. However, this understanding and tolerance will also need to be informed by an appreciation of the global power and significance of racism and its manifestations at all levels of environmental activity.

6

Environmental Monitoring and Planning for Sustainability

DUNCAN BAYLISS and GORDON WALKER

The present debate on environmental sustainability shows a common shared assumption that things cannot go on as they are. How change should be achieved is less agreed. However, many view the pursuit of sustainability in line with the dominant 'technocratic' mode of environmentalism (O'Riordan and Turner 1983) which sees sustainability fundamentally as a technical or managerial concept. From perspectives such as economics or environmental science, sustainability is very broadly seen as being achieved by knowing the state of the environment, forming adequate and effective policy responses and then affecting desired change towards an overall goal.

In this chapter we consider whether such an approach to sustainability is presently viable by evaluating the adequacy of the current environmental monitoring on which our understanding of the present state of the environment is based and from which progress towards future environmental sustainability will be assessed. Environmental monitoring is central to implementing any effective policy for sustainability based on a 'managerial' approach (see below for further discussion). Without an accurate, reliable and appropriate information base, policy formulation will be haphazard and its impacts will not be capable of being monitored. Jacobs observes that, 'without accurate and systematic information about the state of and changes in the environment (which many

Environmental Planning and Sustainability. Edited by S. Buckingham-Hatfield and B. Evans.
© 1996 by John Wiley & Sons Ltd.

countries surprisingly lack) it is impossible to set sustainability targets and to direct policy to meet them' (Jacobs 1991, p. 237). This information base will be built upon the results of monitoring the environment and hence environmental monitoring data quality is of central importance.

Much work to date has assumed that the information base that is needed already exists, or may be relatively easily constructed. Whether these are reasonable assumptions is however open to question. We argue in this chapter that they are overly optimistic, justifying this position by examining the role of monitoring and information in the various approaches being taken to sustainability, the adequacy of the present knowledge base on the environment, the information needs of planning for sustainability and present attempts to meet these, and then assessing the prospects for developing an appropriate information base in the future.

Approaches to sustainability

There are many different approaches being taken towards the pursuit of sustainable development. Any classification is inevitably simplistic but may usefully illustrate the differing strands of thought and ways in which sustainability is being considered.

Environmentalists represented by writers such as Arne Naess (see Naess 1973) and pressure groups such as Greenpeace and Friends of the Earth, have tended first and foremost to emphasise the perceived need for action to stem increasing rates of environmental degradation. One might term this *environmental advocacy*. Influenced strongly by the Club of Rome's pessimistic approach (Meadows *et al.* 1972) whereby it was held that with accelerating rates of change the early manifestations of damaging trends would be difficult to detect and that when they were obvious the damage would be occurring at such a rate as to be probably irreversible, they have urged action now 'before it is too late'. The emphasis has been less on knowing *for certain* about the state of the environment and changes in it and more on seeking action to protect it.

Green *'political ideologists'* approach sustainability by centring mainly on changing *attitudes* to the environment rather than on specific prescriptions as to policy or practical changes that are to be undertaken. Jacobs, observes that,

> some green writing seems implicitly to assume that in a green society environmental sustainability will be achieved because people's attitudes and motivations will have changed: they will be non-competitive and non-materialistic and 'in harmony with nature'. Some writers have gone so far as to say that sustainability cannot be achieved until such a transformation has occurred. (Jacobs 1991, p. xviii.)

Environmental *'managerialists'* approach sustainability very differently being based in research traditions such as economics and environmental science which are used to measuring in some way the phenomena they deal with. O'Riordan and Turner (1983, p. 9) comment that: 'the unbridled technocentrist is dying, a new variety of environmental manager is emerging, one who seeks to compromise (not always to a middle ground) between the demands of creating more wealth and the need to safeguard against risk, environmental damage.' This 'cautious reform' which they suggested would dominate in the short term, has been very active in the rise of sustainable development as a policy goal, providing the broadly accepted basis on which progress is to be made. For this reason a more detailed analysis is needed.

Pearce *et al.* (1989) popularised the managerialist approach to sustainability from the perspective of economics, with their demands for environmental data to be incorporated into national economic accounting. Probably the central issue in the way that they tackle sustainability is in trying to find a means to make the ideal of inter-generational equity operational. In essence the approach they support requires appraising and knowing about the state of the environment, our environmental assets, and then seeking to ensure that the total stock of these is not diminished between generations. The two main approaches to measuring the environmental assets involve ensuring that either (a) the total stock of natural and artificial assets, or (b) the total stock of natural assets, is not diminished. The former approach assumes that there may be substitution between artificial and natural assets. Whilst this may well be held to be true to some extent, Pearce *et al.* (1989) come down on the side of the second approach on the grounds of the level of uncertainty we face in managing for multiple possibilities in the future, arguing that, 'Each generation should at least inherit a similar natural environment' (Pearce *et al.* 1989, p. 37).

This approach is a form of 'constant stock rule'. Thus the environment may be, 'likened to a stock of natural capital yielding a flow of services to the economic system (i.e. its essential economic functions), then sustainable development of that system involves maximising the net benefits of economic development, subject to maintaining the services and quality of the stock of natural resources' (Pearce *et al.* 1989, p. 42).

In practice, trade-offs will be sought with depletion in one area to be compensated for by gains elsewhere, although the extent to which this may be achieved by artificial capital is debatable. Yet, however one interprets this approach, it implies knowing a lot about the stocks, flows and balances of environmental assets, to be implemented meaningfully. Recent interest in identifying 'sustainability indicators' of the natural environment to ensure that minimum stocks are kept and that critical thresholds are identified and not crossed arises from this rationale (Steer and Lutz 1993). A key weakness in the work of Pearce *et al.* (1989) and others, is the lack of attention to the issue of whether such indicators and an information base on the environment, of utility to planning for sustainability, can in fact be found.

The 'managerial' approach and information needs

The managerial approach to sustainability that has been outlined above, is essentially based on a positivist or scientific conception of the environment and its management. The primary characteristic of this approach to conceiving the environment, as already noted, is that much information about the environment is sought to inform policy and decision-making. The assumption is clearly made that the environment is objectively measurable and that measurement will enable modelling of its processes, which must be understood if the environment is to be managed. Thus measurement is considered to be meaningful and the issue of what to measure and how, becomes central. In terms of modelling the environment, systems theory has had a major impact, with talk of stocks and flows by Blowers (1993) implicitly recognising this analytical approach.

Measurement and modelling of the environment may potentially establish the present state of the environment and allow change from that baseline to be measured. This is essentially an application of a form of the scientific method variously termed positivism or the hypothetico-deductive method (Simmie 1993).

What then does this approach to sustainability require us to know about the environment? What monitoring does it need? We need to know which aspects of current activities are not sustainable and what impact any policies adopted have in moving towards a preferred course of development. The implication for environmental monitoring, given the holistic nature of sustainable approaches to environmental management, is that a comprehensive set of information on many diverse aspects of environmental quality and resource use is required (Friend and Rapport 1991). For each of these a baseline of some kind needs to be established from which change can be measured.

This need to measure sustainability applies at whatever level it is being pursued, from global down to local (Norgaard 1988). At a global level an aggregation of, for example, total emissions of greenhouse gases and how these are changing is needed to feed into modelling of global warming processes and the construction of international agreements on cuts and emissions. At an international level consistent and comparable data are needed to, for example, inform the targeting of European Community (EC) resources at areas within the Community where environmental problems are more severe, or to judge comparative contributions to joint targets. The EC committed itself to producing its first EC-wide State of the Environment Report by 1996, partly on this basis (CEC 1994a). At a national level, comprehensive and meaningful environmental information is needed to feed into the kind of green national accounts advocated by Pearce *et al.* (1989). At a local and regional level, monitoring information is needed to guide and direct practical action for environmental management and improvement.

Recent developments in environmental information

A range of responses have come about as a result of a recognition of this need to have environmental information available to guide environmental policy and the pursuit of sustainability. These include the development of state-of-the-environment reporting (SoE reporting) and the production of indicators of sustainable development. Each of these will be commented on briefly as regards their approach to knowing about the environment and how they may inform sustainability policy development.

State-of-the-environment reporting has developed in the last 25 years with reports being undertaken at various spatial scales from international to local. Japan and the USA were the first to start national reporting in 1969 and 1970 respectively (Healy 1987). In the UK the first national SoE was produced in 1992 by the Department of the Environment (DOE 1992) although local-scale reports had been undertaken earlier, the first by Kirklees Metropolitan Borough Council in 1989 (Kirklees Borough Council 1989). SoEs are essentially collections of available environmental data put into one document for a given spatial area. Whilst they are valuable as easy sources of information, and may play important roles in helping public education and awareness, their wider usefulness is open to question and is not at all clearly established by those that have produced them. Whilst they are often assumed to provide a valuable policy role and an aid to sustainability, the collection of available data does not necessarily tell one much about the state of the environment *as a whole*, which is a matter of concern to planning for sustainability.

The production of environmental indicators similarly represents an attempt to use existing data in a more effective manner. Here the goal has been to produce a measure (or set of measures) of environmental quality or environmental performance which may be reported and traced over time, much as measures of inflation or unemployment are currently employed. This possibility of developing indicators was enthusiastically advocated in the 1970s by Inhaber (1975), who argued that they should form the basis of action by governments, anti-pollution groups, industry and the public. However, little effort was directed to this end until more recently when a number of initiatives have been taken and proposals put forward. For example, the Organization for Economic Co-operation and Development (OECD) attempted to compile a set of preliminary indicators in a report issued in 1991, intended to provide the basis for comparing environmental performance amongst member countries, enabling the integration of environmental and economic decision-making and aiding communication with the public. Their list comprised 18 'environmental' indicators covering both the quality of the environment itself and certain national and international goals and issues (OECD 1991). An example of an index combining together a series of indicators is that proposed by Hope *et al.* (1992) for the UK. This is based upon the combination of nine parameters including the number of oil spills, fertiliser deliveries to agricultural use and number of new dwellings started (some criticisms of this selection of parameters and the basis of the index are made later).

Common to both SoE reporting and the production of environmental indicators is a dependence on environmental monitoring. Each will only be as good as the results of the monitoring it draws on allow it to be, although this is rarely recognised. Lang (1979), in an early state-of-the-environment reporting exercise in the USA, noted that there was a tendency to assume that 'the data are there' when in fact often they were not. Several critiques of more recent SoE reporting also note that they tend not to be either explicit about the quality of these data and the monitoring that produced them, or the manipulations and analysis the data are put through (Healy 1991; Bayliss 1994b). The environmental indicators recently constructed as discussed above are also largely uncritical of data quality. The OECD report at least recognises problems of data quality and at various points issues warnings about making comparisons, but the extent of incompatibilities revealed by the technical annex is never fully acknowledged and fundamentally undermines many of the tables and graphs that are presented.

Evaluating current environmental monitoring

Having identified environmental monitoring as a central issue, it is worth considering the data quality issues in more detail. 'Data and information quality' has been referred to in a very general sense thus far. It is taken here to mean a variety of things more specifically. It means that the actual act of measurement of the environment is undertaken in a scientifically acceptable manner – although individual measurements on the environment do not automatically constitute information. Measurements must be undertaken in a comparable manner and then be aggregated in a way that is *spatially* and *temporally* representative in order to identify and probably to summarise the state of, and trends in, environmental parameters. The need for comparability and the need to summarise arises within any given monitoring regime such as monitoring across a drainage basin, and in making comparisons year on year (over time) and between places (over space). Thus the results of monitoring are processed to result in what constitutes information on the environment.

The complexity of this process and the potential for value judgements and choices and incomparability of methods to affect the information which is presented at the end of the process is considerable. This leads to the criticisms of state-of-the-environment reporting and the development of indicators of sustainable development already mentioned. If they pay scant attention to these data and information quality issues then the policy recommendations which flow from them are entirely open to question – since the picture one has of the environment affects the policy prescriptions one makes.

It would clearly be a major task to present a thorough evaluation of the environmental monitoring undertaken across a full range of aspects of the environment in different countries, and one which is far beyond the scope of this chapter. However, it is possible to illustrate some of the limitations and problems that do exist with current monitoring practices.

At a general level, a lack of monitoring results, and also difficulties with specific aspects of monitoring have been noted in various places. In the UK, environmental monitoring has accumulated over time in response to various perceived problems and pressures, such that its contemporary pattern far from reflects any systematic assessment of information needs. For example, air monitoring in the UK is dominated by the measurement of sulphur dioxide and smoke levels which are both largely historic forms of pollution in long-term decline. More 'modern' pollutants such as nitrous oxide and ozone which are under less control and which show increasing trends are monitored far more sparsely. Such observations have led the UK Quality of Urban Air Review Group to argue that the

DOEs 'total monitoring effort is now in urgent need of rationalisation and review' (Department of the Environment 1992c). The UK water monitoring network has been similarly criticised by the Royal Commission on Environmental Pollution, with a recent report calling for more detailed monitoring of a greater range of pollutants in surface waters, routine biological monitoring and the setting up of a national groundwater monitoring network (Royal Commission on Environmental Pollution 1992).

At an international level, data problems multiply to an extent due to problems of incompatibility between different countries (although these can also be seen between different places in a single country). The EC Fifth Action Programme on the Environment 'Towards Sustainability', briefly acknowledges these problems, and calls for: 'higher quality environmental data in greater quantities, gathered and interpreted in a standardised manner by designated bodies in each member state' (CEC 1992a).

The Brundtland Commission similarly commented on the 'limited international capability for monitoring, collecting and combining basic and comparable data needed for authoritative overviews of key environmental issues and trends'(WCED 1987).

Whilst such statements at least recognise there is a problem to be addressed, they are not accompanied by any realistic assessment of the extent of the problem or of the difficulties to be faced in improving matters. Biodiversity is a key area of policy for sustainable development yet there is no consensus available as to how this can be defined and therefore measured (Bishop 1993). It can be defined in respect of genetic diversity, species diversity or ecosystem diversity, and for each of these there is no clear or straightforward measure that can be made. Categorisations, for example, of species at risk and of protected areas can vary considerably between different countries and there is much scientific uncertainty involved, so that the practicality of making any meaningful measures of biodiversity must be fundamentally called into question (Hohl and Tisdell 1993).

The OECD (1991) attempt to produce internationally comparable environmental indicators (discussed earlier) displays similar problems although of a less fundamental nature. Virtually none of the data collected together for each member country were entirely comparable, the technical annex to the report revealing, for example, that information related to different years, and that data for some countries were only crudely estimated. For river water quality, information is only presented for one or two rivers per country, which cannot be at all 'representative', and in some cases, for example with hazardous waste, definitions of what is being measured vary enormously. Furthermore the report provides no detail on monitoring methodologies or sampling frameworks which are essential to achieving any degree of meaningful comparison.

These more detailed comparability problems were revealed in a study examining the environmental monitoring occurring at the local level in four European cities: Erlangen (Germany), Rennes (France), Stoke-on-Trent (England) and Thessaloniki (Greece) (Bayliss *et al.* 1993). This considered a range of environmental parameters related to air, water and land quality. In each city an agreed set of data was sought, together with information on how these data were collected in the first place, the uses they are put to, and the arrangements for access by third parties to them.

Table 6.1 shows the results of some of these comparisons for a number of forms of pollution. Clear differences between the monitoring practices in each city are apparent. In a few cases there are data gaps as no monitoring is taking place. In the case of surface water in Thessaloniki, this is explained by the lack of surface flow for most of the year, but in other cases such as land contamination in Rennes, there is no clear physical reason for the information gap. When monitoring *is* taking place there are significant differences in the sampling frameworks being used. There is much inconsistency in both the temporal and spatial aspects of sampling programmes with air pollution, radiation, surface water and to an extent drinking water demonstrating marked differences. The results of monitoring are also summarised and reported in different ways so that for surface water France and the UK have five classes of river quality, although these are differently defined, whilst Germany has seven.

A more general point is that the study found that it was consistently difficult to obtain full details on monitoring methodologies employed. This was at least as much due to lack of readily

Table 6.1   Summary of monitoring information for four cities

Type of pollution	Thessalonikii	Stoke-on-Trent	Erlangen	Rennes
Ambient air quality	3 sites 5 pollutants	4 sites 2 pollutants	3 sites 6 pollutants	2 sites 5 pollutants
Radioactivity (Gamma)	Occasional surveys	6 regular sites + random sites (weekly)	25 sites (quarterly)	1 site (continuous)
Surface water	No monitoring	30 sampling points	1 sampling point	5 sampling points
Ground water	One-off survey of 22 wells	1 borehole	21 + sampling points	No monitoring
Drinking water	At least annual sampling of central well	Frequent sampling by two bodies	Frequent sampling	Very frequent sampling by two bodies
Land contamination	Survey of background levels only	Ad hoc, limited monitoring of a few sites	Survey of background levels and thorough site investigations	No survey undertaken

available documentation as to unwillingness to divulge such information to third parties. The result of this, however, is that it is difficult to check the quality of data upon which environmental reporting is based.

Whilst this study related to only four cities it clearly indicates that there remains considerable work to be done before one may reliably construct a rigorous, comparable and holistic understanding of the environment at the local scale. Given this, then, the aggregation of such data to higher orders (for example, national or international) will simply introduce further potential sources of error or bias in the information. If there are inconsistencies in data gathering at a local level, these will multiply or be aggravated by aggregation, as the basis upon which the data were collected and the variability in the data at the local level is lost.

Various attempts have been made to establish a greater level of international compatibility in how data are collected, but their limited success further demonstrates the difficulties involved. For example, the Global Environmental Monitoring System (GEMS), established by the UN Environment Programme (UNEP) and developed through the 1980s, aimed to provide an integrated network of monitoring points around the world, monitoring the background levels of certain 'key' contaminants such as pesticides, sulphur and mercury compounds (Port 1980; Rovinsky 1982). Whilst significant progress has been made, it has proved very difficult to achieve extensive international coverage and to ensure the reliability of monitoring practices. Data from the system, for example for air pollution, show a patchy pattern of selective monitoring with many gaps from year to year. At a European level the possibility of achieving more compatible monitoring and reporting of surface water quality has been explored. However, a recent symposium (Walley and Judd 1993) revealed that whilst harmonisation of such matters as sampling and analytical methods, the selection of monitoring sites, and biological classification systems was widely supported in principle, there are many practical difficulties to be addressed which for some make harmonisation only a long-term objective, and for others an almost impossible goal.

Outstanding issues

Following from our analysis of present environmental monitoring and how the quality of this is central to a scientific approach to the environment, some significant outstanding problems remain.

Inherent problems in a positivist/scientific approach to sustainability

The whole approach of measuring environmental parameters to inform environmental and economic policy and planning is likely to come up against the perennial problem of the scientific enterprise: namely that science has had a consistent thrust away from a single integrated body of knowledge to an apparently ever-increasing fragmentation of our understanding of the world about us. Science has resulted in our knowing more about increasingly highly specialised areas, whilst the integration of this knowledge remains a growing challenge (Chalmers 1978). This trend is reflected in post-modern philosophy talking about the fragmentation of consciousness (Lyotard 1982). To apply the concept to monitoring the environment, by analogy, one may think of science as having yielded a very large number of jigsaw puzzle pieces, with varying amounts of detail on each, whereby one has no idea from the individual pieces what the overall pattern should look like when the pieces are assembled. The matter is not helped by the fact that different people suggest putting the same pieces together in different patterns; the results of the same monitoring may be used to argue for different pictures of the state of the environment and then different prescriptions as to what to do to manage it. The problem here for planning for sustainability is that if it is pursued in the way outlined by Pearce *et al.* (1989) and others, then we need to integrate disparate aspects of knowledge about the environment in some way so as to allow an overall assessment of our impacts on it (which is implicit, for instance, in the concept of ensuring a constant stock of environmental assets).

To get around this impasse one could theoretically develop immensely sophisticated models of natural systems derived from scientific observation, which would allow some measure of prediction of the effects of particular actions by humans on the environment, which would in turn inform normative theory for planning. However, there are several difficulties with this. First, as already shown, we do not as yet have that level of understanding and any models produced may prove unworkably complex. Second, action is being urged to be taken now, according to the so-called 'precautionary principle' (O'Riordan and Cameron 1994) whereby it is argued that if the potential consequences of inaction are unjustifiable then action should be taken before there is scientific proof to justify it. Third, experience to date of attempts to be rationally comprehensive in decision-making in planning has not been encouraging (McConnell 1981; Simmie 1993).

Attempts to borrow directly the methods of analysis and construction of knowledge of the physical sciences where the positivist scientific approach originated, for use in the social sciences, such as planning (in its broadest sense and land use planning in particular) have been heavily criticised. McConnell's (1981) main argument against comprehensive approaches to planning is that theory without being spatially, temporally and client group specific, cannot be falsified, leaving a lack of any basis to establish the reliability of the claims it makes to producing reliable knowledge.

One of the key weaknesses of much discussion about sustainability as already outlined has been the implicit assumption that this information and understanding gap can be bridged. There is the danger that sustainable development may be sidelined from affecting the fundamental change its advocates want unless clear analyses can be made of environmental degradation and potential ameliorative actions within the positivist/scientific epistemology which has dominated environmental management in the developed countries to date. It is thus an important research issue to identify the dimensions of such an approach and to establish whether the environment in its entirety can be conceived in a way which makes this managerial approach feasible.

The quasi-scientific use of environmental monitoring data

There are a growing number of examples of environmental data being used quite unquestioningly. Even if the data are good they must still be used appropriately. An interesting example of the danger of assuming the validity and quality of data and then using it uncritically or indeed somewhat naively, is the environmental index proposed by Hope et al. (1992) for the UK. This displayed a lack of clear and critical thinking in the way they intended data to be used and combined (Bayliss and Walker 1992). For example, a number of the nine parameters selected showed little or no clear relation to environmental quality or impacts upon it – simply using the 'number of oil spills a month' as an indicator neglects the size and location of these spills which are clearly vital determinants of environmental impacts. The various components of the index are weighted by public perceptions of their significance, but a measure of perception taken at one point in time is used to weight a data series over a period in which perceptions may change. The weightings chosen will also have significant effects when the various components of the index move in different directions, which is quite likely given that, unlike the retail price index, the components which make up the environmental index do not have a high degree of interconnectivity.

The problem, though, is not simply one of improving the quality of environmental monitoring programmes and the methods of deriving information from them, because any given picture of the environment does not necessarily lead simply and directly to 'obvious' actions to manage it. Substantive theory does not automatically validate normative theory; or to put it another way, description of the environment does not automatically validate the prescriptions one makes for its management. This leaves a considerable problem, where prescriptions for planning (derived from the axioms of sustainability drawing on description of the environment) beg to be tested, but such testing is hard to propose. The sheer complexity of the environment and the timescales of environmental degradation make it hard to 'experiment' on the environment in relation to many potential human impacts to derive the best course of action, even assuming such testing is possible to specify and politically possible to execute. Just because the evidence used to describe the state of the environment was collected in a rigorous scientific manner does not mean that normative theory built from it is similarly testable. Much other normative theory used in relation to planning (such as Marxist analysis) has proved hard to test too. In practice, this issue seems hardly to have been touched upon within the literature on sustainability and is largely side-stepped by many environmental advocates. They often simply urge action *now* on the grounds of the precautionary principle as outlined earlier. The irony here is that scientifically derived measurement and monitoring of the environment is yielding data that are used to construct a picture of the environment which is in turn used as justification for immediate action. Yet, when the scientific approach meets problems as outlined here, it is seen as something to be side-stepped.

Monitoring and alternative approaches to sustainability

Earlier in this discussion we outlined various alternatives to the managerial approach to sustainability that is currently dominant and that we have argued is based upon a positivist tradition. Given the difficulties we have presented in collecting the information necessary to support this approach, one might reasonably ask what the environmental monitoring implications of alternative approaches to sustainability might be. Whilst it may be possible, as the green 'political ideologists' have argued, first and foremost to seek a change in attitudes rather than try and plan and sustainably manage the environment, monitoring is still a necessary activity. This is because, first, such an approach has been shaped and informed by the results of environmental monitoring; second, the mechanisms by which change will result are rarely explicitly stated; and third, any resultant change may not be known about unless the environment is measured and monitored in some way. Therefore the issues raised here relating to environmental monitoring are likely to remain significant however one approaches sustainability. However, if the limitations of monitoring are sufficiently fundamental in nature, then the scientific/positivist approach to pursuing sustainable development is clearly undermined far more than any other.

Conclusion

It follows from the arguments we have pursued that far more attention needs to be given to data collection and data quality issues within policy developments related to planning for sustainability. There are many sources of variability and hence uncertainty inherent with much information on the state of the environment. This arises from the way it is collected, analysed, presented and interpreted, potentially preventing meaningful comparison between data at different spatial scales and locations. There are therefore real dangers in taking current environmental monitoring and using its results as given, and careful questions need to be asked about the quality of the data currently available and being presented in an increasing number of compilations and reports. It may well also be premature to talk about feeding environmental data into 'green' national accounts, until the basis of those data is more assured; in this respect there is some discussion of data problems in the 1993 'Blueprint' report (Pearce *et al.* 1993) although the viability of measuring sustainability is not at all challenged. As a consequence, data 'health warnings' need to be more explicit and more visible, so that they are less easily overlooked, ignored or manipulated in the rush for information to feed into policies and information programmes.

There are a number of reasons for this current lack of rigorous attention to data quality issues. Many commentators on sustainable development have failed to be explicit about the conceptual approach they are taking to knowing about the environment and the information needs of the approach they advocate. If one's approach relies on attempting to know objectively about the state of, and changes, in the environment, then the quality of the environmental monitoring which informs this approach is of paramount importance. In addition, many of those commenting on sustainable development have little or no knowledge of physical environmental monitoring. Similarly, those most involved in the development of monitoring methodologies and techniques are typically scientists with highly specialised areas of interest with often limited understanding of policy development and implementation. One of two traps is consequently fallen into: the policy-maker assuming, without checking, that the information is there; or the scientist assuming, without tracing the linkages, that policy is based on results of thorough and rigorous environmental monitoring.

It seems likely that despite any fundamental questioning of the entire basis by which we go about knowing about the environment and then attempting to manage it, sustainability will continue to be advocated. However, if more and more radical changes are proposed or enforced on people's activity and behaviour then such fundamental questions may be thrust to the fore.

7

The Changing Face of Environmental Policy and Practice in Britain

JULIAN AGYEMAN and BEN TUXWORTH

Environmental legislation, policy-making and practice in Britain is in a state of flux. It is characterised by conflicting approaches from a range of different players. The departmental structure of central government reflects the more traditional concerns of a national administration, into which policies for environmental protection and sustainable development are integrated with difficulty. The Department of the Environment is the governmental lead body on environmental policy, but other government departments administer policies with just as much if not more influence on the state of the environment. Thus the Ministry of Agriculture, Fisheries and Food administers agricultural support schemes; the Department of Transport is responsible for vehicular pollution control and the roads construction programme; the Department of National Heritage cares for ancient monuments and the listing of historic buildings, and the Department of Trade and Industry's responsibilities include power generation.

Environmental policy increasingly impinges on the work of all government departments, but the tendency in the UK has been to treat environment as an add-on to existing policy, rather than as requiring a strategic response. The government's 1990 White Paper

Environmental Planning and Sustainability. Edited by S. Buckingham-Hatfield and B. Evans.
© 1996 by John Wiley & Sons Ltd.

'*This Common Inheritance*' and its subsequent annual reports (DOE 1990, 1992a, 1993c, 1994b) present environmental policy as an integrated approach across the board, but the lack of any commitment to targets in environmental quality, and the continued emphasis on voluntarism and the role of market forces means that it can still be argued that central government has no coherent environmental strategy or policy, just a set of reactive approaches to given issues. This lack of integration of policy results in a situation of conflict within government, where despite ubiquitous 'Green Ministers', some departments are only just beginning to accept the need to reassess their own objectives in the light of environmental issues.

Two factors lie at the heart of governmental reactivity in environmental policy-making: the tradition of voluntarism in environmental regulation in the UK, which produces a *laissez-faire* attitude to increasingly serious issues of environmental protection; and the ideological commitment to non-interventionism, privatisation and deregulation that constituted the major touchstone of the Thatcher administration and which continues to dominate Conservative policy in the 1990s. It is these factors which explain the paralysis and vulnerability of the government in the face of the emerging international environmental agenda with its focus on explicit and consensual planning for sustainable development.

It is also these factors which have left Whitehall apparently bemused by the growing confidence in local authorities who are responding with vigour to the new environmental agenda. With the initiative not in central government hands, the contrasting approach of local and European tiers of government bolsters the critique of national policy developed by a vast network of environmental organisations, protest groups and direct action specialists who are better informed, networked, resourced and more media-wise than ever before.

In this chapter, we argue that the rapidly developing international and European environmental policy agenda forms the greatest single influence on environmental policy and practice in Britain today, at both local and national scales. Like other nations in Europe, the UK is required to adopt or enforce EU Directives, regulations and decisions. Cynics argue that it is to this obligation that we owe the small amount of environmental legislation enacted in the last two decades. Whilst the motivation for a piece of legislation is not always clear, it certainly seems likely that it was compliance with EU Directive 79/409 on Wild Birds, which forced the introduction in the UK of the Wildlife and Countryside Act (1981), the only significant piece of environmental legislation of the Thatcher era. That compliance with EU legislation should be the only effective motivation for domestic environmental protection, despite mounting public concern about the state of the environment throughout the 1980s and 1990s, and increasingly effective, influential and media-worthy campaigning strategies by a range of pressure groups, gives some measure of the significance of the international arena.

Despite this, to the outside observer, European environmental legislation often seems to arrive in Whitehall as an uncomfortable surprise. It sometimes requires expensive remedial action to achieve compliance with new, and invariably higher standards. Such reactivity is exemplified by Britain being the only European country still dumping untreated sewage sludge into the sea. EU Directive 91/271 on Urban Waste Water Treatment sets minimum standards for the treatment of urban waste waters (domestic sewage and industrial effluent) in towns of over 15 000 people, and is to be implemented in stages between 1998 and 2000. The Directive will have a major impact on Britain's current out-of-step 'dilute and disperse' policy approach to untreated sewage in which only 2% of Britain's sewage outfall to the sea receives secondary treatment.

At the same time, the development of environmental policies and strategies in local government is one of the areas of greatest achievement and optimism in the spectrum of local authority activity (Agyeman and Evans 1994), and seems to be offering a new mandate for local democracy. Local authorities have been quick to seize the opportunity, and adopt an approach which puts them in step with all the significant environmental policy agendas, both European and worldwide. Many are now pushing for comprehensive legislation and policy, higher environmental standards and are progressing ideas about local structures for sustainability.

The role of local authorities in statutory environmental management and protection, as well as in policy-making (which in many cases far exceeds statutory requirements), must be seen in the context of the current government's approach, which can be characterised by the rejection of long-term planning; a reliance on market forces; deregulation; public service privatisation and Compulsory Competitive Tendering (CCT); strict public sector spending controls; the erosion of local authority powers; local government reorganisation and the reliance on voluntary or economic controls over regulation. Despite this rather bleak operational framework, local environmental policy-making, including the integration and corporatisation of policies into comprehensive strategies for sustainable development, continues apace.

Much has occurred since the 1989 publication of Friends of the Earth's *Environmental Charter for Local Government* (FOE 1989) and the Local Government Management Board's *Environmental Practice in Local Government* (LGMB 1990, 1992). Action in traditional areas of local government concern means that most authorities now have programmes in place which include waste reduction, composting and recycling initiatives, the development of cycle lanes, nature and energy conservation policies, and more sound purchasing practices. Beyond these specifics they are also beginning to grasp the nettle of their overall environmental performance, what it means in management terms, and their changing role in the face of a rapidly evolving approach to sustainable development at the local level. Such foresight in local authorities has not gone unnoticed: it led, for example, to Leicester City Council being presented with a Global 500 Award by the United Nations. This to some extent embarrassed central government into taking up the environmental gauntlet thrown down by local authorities at the Rio Summit.

In seizing this agenda, local authorities have found new allies in environmental organisations, who, after a period of mutual mistrust, have come to recognise local government as a local environment protection agency, increasingly willing and able to perform such a role. This organisational symbiosis will develop and strengthen as a result of each party increasingly 'leapfrogging' the national tier and developing direct links with European environmental decision-making fora. The same process of taking an international policy initiative and bringing it straight to the local level can be seen in the development of consensual approaches to environmental issues in the Local Agenda 21 process.

Apart from the boost of local authority involvement in the Rio Summit, there are a variety of reasons for the range of environmental innovation at a local level. These include the following:

- the allocation of responsibility for environmental policy work, often through the appointment of corporate environmental co-ordinators in most authorities;
- the support networks offered by the local authority associations and the LGMB, together with the international associations such as the International Council for Local Environmental Initiatives (ICLEI) and the International Union of Local Authorities (IULA);
- new confidence in the role of local government gained from the principle of subsidiarity;
- the EU Fifth Environmental Action Programme which restated a pivotal role for local authorities;
- increased public awareness and greater influence of environmental organisations;
- better anticipation of EU environmental thinking as a result of new links with Europe and the ability to deal directly with Directorate General XI.

The international context: influencing national and local agendas

There is insufficient space here to review in detail the impact of the numerous international and European conferences, working groups, commissions, committees and research studies which make up the international environmental policy agenda. However, three stand out as landmarks in terms of their influence on British national, and local environmental agenda setting and policy-making:

- the 1987 report by the World Commission on Environment and Development ('The Brundtland Report') entitled *Our Common Future* (WCED 1987);
- the 1992 United Nations Conference on Environment and Development (UNCED or 'The Earth Summit') (UNCED 1992);
- the EU Fifth Environmental Action Programme (1993–2000): 'Towards Sustainability' (CEC 1992a).

Despite their quite different origins, all of these initiatives put forward the concepts of sustainability and sustainable development as the means and end for environmental programmes into the next century. While a universally accepted definition of sustainable development remains elusive, and the term is often used interchangeably with 'green' or 'environmentally friendly', the language of sustainability offers a much sounder conceptual basis for environmental theoreticians and practitioners. This is largely because it integrates concerns about the fate of the planet with issues of economy, society and democracy, and offers a basis for action for all stakeholders in the environment. The concept of sustainability also provides a framework for detecting and measuring progress towards, or away from, a desired state. This has spawned useful new management tools in environmental policy-making and practice, such as 'sustainability indicators', 'targets', 'monitoring', 'sustainable budgeting', 'strategic environmental assessment', 'environmental capacity' and 'reporting', amongst others.

These influences are still comparatively new. As a result, the attitudes of those responsible for their implementation in Britain, and the policy agendas they control, are still in the process of being reorientated from 'green', towards the more measurable goals of sustainability. The extent to which these global ideas are being incorporated and delivered by national and local policy frameworks is still a matter of some conjecture. Contrasts between the demanding policy and programme undertakings of many local authorities (see for example statements from East Hampshire District Council (1993) and the London Borough of Hackney (1994)) and the bland assertions of the government's 'Strategy for Sustainable Development: the UK Strategy' (HMSO 1994a) are nevertheless dramatic. They leave the reader in little doubt as to who is leading the development of partnership approaches to creating sustainable communities, life- and work-styles.

The national context

The need to anticipate and plan for national wants and needs after the 1939–1945 war heralded a new proactive approach in environmental policy-making and legislation in the UK. Where environmental legislation had previously been concerned primarily with standards of protection, the Town and Country Planning Act (1947) introduced the world's first integrated planning system and enabled a strategic approach to the allocation of environmental costs and benefits. The British system is unique in that it incorporates controls over land use, as well as the design and form of built environments. Much of the 1947 Act still pertains today, although the way the planning system works has in practice been radically altered. Much of the change has resulted from a rejection by Conservative administrations of the socio-economic remit of planning. Instead they favour a deregulated, minimalist, pro-development approach where the market will distribute resources, and where the role of planning is reduced to a form of development control.

The land use planning system nevertheless remains central to environmental protection in the UK. After the questionable success of 1980s experiments in allowing the market to determine the form and function of developments (consider the history of London Docklands), plan-led planning has seen a renaissance, a response in part to the international environmental agenda. The Planning and Compensation Act (1991) has brought the system of planning into the mainstream of environmental policy. The Planning Policy Guidance note series has been rewritten, with sustainable development principles advocated throughout. The PPG series nevertheless gives little in the way of practical guidance on how sustainable development might be applied in the local context. The most recent guidance to date, PPG 13, seeks to integrate land use and transport planning, and acknowledges that 'continual growth in road transport and consequent environmental impacts present a major challenge to the objective of sustainable development. Traffic growth on the scale projected could threaten our ability to meet objectives for greenhouse gas emissions, for air quality and for the protection of landscape and habitats' (DOE/DTp 1994). The consequences of such guidance are discussed in our cycling case study.

Given the global and European endorsement of the crucial role of planning in delivering sustainability, it seems inevitable that the future of the British planning system and profession will lie in its response to this new agenda. The profession has been slow to respond in terms of the content of its training regimes, or the extent of guidance issued to planners on the practicalities of introducing the principles of sustainability to the planning process. The Town and Country Planning Association has produced an introductory text on the subject (Blowers 1993), but the Royal Town Planning Institute seems instead to have concentrated on appeals to the DOE to produce definitive guidance. The groundwork whereby a good practice guide might be produced has begun as part of the Department's research programme, but the guide is unlikely to appear for at least another two years at the time of writing (March 1995). In the meantime it has been left to the Council for the Protection of Rural England (1993) and English Nature (1994) to interpret sustainable development to planners. The Countryside Commission, English Nature and English Heritage are also planning joint guidance to local authorities as to the role of their particular interests in the Local Agenda 21 process; this will appear in late 1995.

It was not until September 1990 that the government produced the first ever 'comprehensive' review of environmental policy in Britain 'from the street corner to the stratosphere, from human health to endangered species' (DOE 1990). *This Common Inheritance* was, in effect, the first step towards integrated environmental policy-making at a national level in Britain. It outlined the environmental challenges facing society, our roles as stewards, the polluter pays and precautionary principles, the need to raise public awareness, the need for international co-operation and a combination of regulatory (legal) and voluntary (fiscal) approaches to policy, veering heavily towards the latter, especially in the second anniversary report. These economic instruments, mentioned in Annex A to the White Paper, might include specific charges, subsidies, deposit/refund schemes, pollution credits and enforcement incentives. What was significant about *This Common Inheritance* was not the policies themselves (few were new), but the indication of commitment to a broader environmental management and protection framework for the future, including better institutional co-ordination.

The Environmental Protection Act (EPA) 1990 brought a similar, integrated approach to legislation. It combined disparate existing provision on air pollution, waste management and disposal, integrated controls over the most potentially polluting processes, litter, the environmental impact of genetically modified organisms and noise and statutory nuisances. As we have observed, its introduction was forced by European environmental legislation. The Act paves the way for the creation of a unified environment protection agency, at the time of writing scheduled for 1995/6, taking on the roles of HMIP, NRA, the Drinking Water Inspectorate and the control functions of various other agencies. But the Environmental Protection Act is not a full code of environmental practice, and the Agency it will establish is not an environment protection agency along US lines. In addition, many issues, such as mobile sources of pollution (e.g. traffic) are dealt with by other pieces of legislation. In October 1994, the original vision for an environmental protection agency was further weakened by Treasury-induced cuts.

The launch of the Green Paper of 'Sustainable Development: the UK Strategy' caused immediate tensions within Whitehall. Amongst other problems, two transport ministers (Lord Caithness and John MacGregor) showed no understanding of the role of transport within sustainable development, nor the concept of demand management, which was beginning to filter into governmental thinking at that time. The final version of the Strategy (HMSO 1994a) shed more light on the drafting process than the road to sustainability. The rejection of the word 'plan' in the title set the tone for what is in effect a restatement of the government's conviction that economic growth is the holy grail and is in fact the meaning of sustainable development. Its critics, and they were many and vocal, claimed the strategy had very little new to say. They pointed particularly to the absence of any vision of a sustainable future, and the lack of commitment to new legislation, regulation, or fiscal measures of any substance with which to achieve it.

The document picks up some of the partnership rhetoric that characterises the literature of sustainable development, but in the main is a continuation of the well-established 'doctrine of voluntarism' in environmental management and protection. This is in marked contrast to the level of commitment made by the equivalent plans for Denmark, the Netherlands and Canada. Apologists for the Strategy point to the fact that it recognises that to continue as we are is not enough. In addition, they note that to bring so many policy interests together in a single document represents a major advance in government thinking.

Current Trends

Unlike countries with a unified environmental protection agency, Britain relies for much of its national environmental responsibilities on a range of autonomous, or semi-autonomous agencies. In addition to the agencies with policy responsibilities mentioned at the beginning of this chapter (often with separate counterparts in Scotland and Wales), there is a range of regulatory agencies, or quangos, with varying degrees of autonomy, such as Her Majesty's Inspectorate of Pollution, the National Rivers Authority, the Health and Safety Executive, the Nuclear Installations Inspectorate, the Countryside Commission, the Forestry Commission, and English Nature. The total budget allocation to quangos now exceeds that to local authorities. Accountability in these agencies is largely to a minister and audit commissioner, yet such agencies are being encouraged to gain even greater independence from government through charging for certain services (the NRA raises over 75% of its money in this way). What effect will even greater autonomy of unelected quangos bring upon the direction and quality of environmental policy-making?

Privatisation and deregulation also have an important role to play in shaping the future of environmental protection. Electricity, gas, water and telecommunications, between them employing huge numbers of people and with vast real and potential impacts on the environment, have all passed from public to private ownership, as has public transport in many metropolitan centres. A private company has no particular mission other than to continue to exist and to be profitable, unlike publicly owned utilities and services, where quite different objectives may be set. One result is that the culture of environmentalism in private sector organisations, where environmental protection is justified only as far as it enhances profitability, must now extend to the privatised utilities. The effect of this in terms of environmental impacts is still emerging. Obviously, the activities of private companies are subject to some regulation, but the climate of voluntarism in the UK leaves much in the hands of businesses to regulate their own affairs.

Efforts have been made by some government departments to put the business case for environmentalism to the private sector, with some success, notably in the creation of the Advisory Committee on Business and the Environment (ACBE). This work has also been taken up by some NGOs, with Business in the Environment and the CBI offering advice and case-study material to private companies on the need for preparedness and action with regard to environmental issues. But the voluntary approach inevitably has its limitations. In the end, most companies admit that they will not comply with environmental actions which require (for example) major capital investment or major restructuring of their operations until it is called for through legislation.

This highlights another issue on which the UK government is at odds with its European partners. Experience in Germany suggests that, rather than curtailing business opportunities, introducing high environmental standards and regulations which actually exceed EU requirements has introduced a climate of innovation and technical excellence. This has resulted in Germany developing an advanced environmental technology industry which now exports to the rest of the world.

As we have suggested earlier, local authorities are now particularly active in developing policies and programmes for sustainable development. The Local Government Management Board/ Environment Resource and Information Centre database of Local Authority Environment Co-ordinators reveals that as of December 1994, 496 of the 542 local authorities in England, Wales, Scotland and Northern Ireland had a named contact on environmental issues. Most authorities now have some form of written environmental policy or charter, and around 50% have environmental action programmes. Other surveys carried out by ERIC on behalf of the LGMB (Tuxworth 1994; Tuxworth and Carpenter 1995) have attempted to assess the extent of the local authority response to the Rio Summit. These reveal that by the end of 1993, around a quarter of all authorities were committed to developing a Local Agenda 21 process, whereas by the end of 1994 the figure had risen to 70% of respondent authorities in a sample of 310. This latter survey also showed that amongst respondents, nearly two-thirds were considering the adoption of an environmental management system, some through the EU Eco-Management and Audit Scheme following the publication of guidance on such schemes by the LGMB in 1994 (LGMB 1994a). More than half had made some progress on state-of-the-environment reporting, and around half had conducted an internal environmental assessment.

It is likely that these figures will continue to rise as the approach proposed by the LGMB-promoted Local Agenda 21 project is refined and the guidance it issues based on a national round-table process (LGMB 1994b,c,d,e,f,g) takes effect. Increasingly, the new vehicle for local environmental policy and practice will be Local Agenda 21. The integration of environmental policy with other policy areas is central to such initiatives, and increasingly local authorities will be integrating more traditional environmental work such as reviews of purchasing, energy, waste, transport and nature conservation, into a strategic approach including environmental assessment and the use of indicators and targets, new stakeholder involvement mechanisms, the development of in-house 'green teams' and regular staff training.

Service provision in local authorities, including environmental services, is increasingly affected by the statutory requirement of CCT. Although this is often seen as a barrier to the introduction of environmental standards, managers are discovering that contracting out can be made to work for their authority's environmental programme. It requires a rigorous assessment of the impact of the service at the time of specification in the contract as once the particular service has been let, it may require a costly variation of contract to effect change. Examples where contracts have an environmental specification currently include waste minimisation objectives in waste collection contracts; ecological and nature conservation management objectives in grounds maintenance contracts; and the specification of environmentally benign cleaning products in cleaning contracts. In addition, with higher environmental standards on the EU horizon, far-sighted local authorities might well consider ways of actually specifying contracts which will be attractive to environmentally sound contractors, including in-house teams, who ought to have benefited from staff training and the like.

Increasingly, local authorities see partnership building as crucial in the delivery of local sustainability. One of the key influences on local authority environmental policy-making and practice has been that of 'people pressure', either as members of local or national environmental organisations (there are more than four million such members) or as individuals. Most authorities realise that they can set up policies, strategies, charters, action plans and structures for local sustainable development, but that these will be virtually worthless unless local people are involved to the extent that they feel a sense of ownership, and empathise with the aims of sustainable development. In turn, this empathy will only develop if the general level of public environmental literacy is raised such that people understand the sacrifices that they will inevitably have to make, and, more importantly, the present, and generational benefits that will accrue. This will have important implications for attitudes to education as a lifelong process, and for a national educational policy (Agyeman and Evans 1994).

The influence of local people in environmental policy-making must not be underestimated, nor must it be taken for granted. It is dependent on the authority setting an example, by getting its own house in order, and requires nurturing through investment and feedback. Investment in a 'community profile' so that authorities know exactly who their constituents are, and regular and appropriate publicity is required if authorities are to meet a major objective of Local Agenda 21 – involving a far wider range of socio-economic and cultural groups than those 'joiners' who regularly attend environmental fora (Agyeman and Tuxworth 1994).

In addition, authorities should see themselves as educators through looking at the extent to which they fulfil the requirements of the EU Directive on Freedom of Access to Environmental Information. This will include an assessment of what information they hold, where it is held, what form it is in, who in the authority will access it upon request and the development of public guidance (Agyeman 1992). Lancashire County Council is using Geographic Information Systems (GIS) to get information on the environment online to schools and libraries. Initiatives such as this could be extended to comply with the EU Directive.

Feedback to local people is an essential part of partnerships for local sustainability. Traditionally, this has been done passively, through newsletters or other media. A more active method which is now gaining ground is 'ecofeedback'. Developed in the Netherlands, and being used by Woking Borough Council, ecofeedback requires householders to keep a record of, for example, their energy consumption. Over time, these data can then be compared to local and national data, allowing the householder to assess his or her performance in relation to preset local and national energy conservation targets. Judging by the interest shown by the public in recycling, ecofeedback schemes could be applied to different issues, and give people a real sense of ownership and primary environmental care.

Whilst the generally upbeat approach to Local Agenda 21 by local authorities is to be welcomed, it would be a mistake to underestimate the substantial problems facing the community-based approach to delivering sustainability and sustainable development in the UK. Greatest of these is the social alienation so intimately linked with the policies of the Thatcher years. The focus on individualism and economic liberalism has produced a situation where, for many people, 'community' does not exist in the sense defined in Agenda 21. Policy responses to this trend include a growing interest in communitarian ideas – the notion that atomised societies have no sense of community or social solidarity, and that this vital social fabric must be rebuilt. Such ideas are now gaining ground in both right and left wing circles. Part of the task for local authorities, in seeking to develop a Local Agenda 21 process, is to translate such thinking into practice.

Other issues are also beginning to emerge. Agenda 21's focus on democratic, participative mechanisms for the delivery of sustainability glosses over a number of the problems which those taking this approach are encountering. Not least of these is the potentially problematic relationship between traditional democratic structures and the new fora, roundtables and groups being formed in the name of local sustainability. Where Agenda 21 calls for the needs and values of all stakeholders, including 'major groups' to be brought under the umbrella of these new structures, Local Agenda 21 processes are responding by convening meetings at which diverse 'representatives' of those groups are invited. But what mandate do those who attend really have? Who do they represent? Other problems arise when the decision-making power of such new structures is considered. If the views expressed and the commitments made at such meetings are to have any value it must be because some executive power has been handed on to the forum or roundtable. In a democratic system the corollary of power must be accountability, and yet it is not at all clear where accountability for a decision made by a local forum would lie.

A case study: trends in cycling policy and practice

It would be hard to find a more contested area of environmental policy than transport, nor a better symbol of sustainability than the bicycle. It is cheap and empowers the individual without impacting on the community. Some 99% of men and 87% of women can ride bicycles (MINTEL 1989). This gives bicycles greater potential than the car as a mode for mass transportation, as only 78% of men and 48% of women have driving licences (Boulter 1994). Bicycles consume little energy in production and present few problems of disposal. They barely pollute in use, offer numerous health benefits, and can be cheap, convenient and enjoyable. They can provide employment opportunities through manufacture and servicing. Those who cycle to work perform better than their sedentary colleagues, and employers encouraging cycling often find problems of lateness and absenteeism much reduced by their introduction (Unwin 1993). Infrastructure for cycling is relatively cheap, and in any case not a prerequisite for increased use given the UK's substantial network of roads.

Some 50.6% of all trips in the UK are made by car. It has been estimated that around 62% of all trips are of less than 5 km. This distance is easily achievable by bicycle, and if the bicycle replaced the car for trips of this distance the percentage of all trips made by bike in the UK would be around 28%, similar to the figure currently achieved in the Netherlands. But in the UK, only 2.5% of trips are made by bicycle, less than a tenth of the Dutch figure (Bannister 1990).

Why do we cycle so little in the UK? Most of the studies that have examined this question conclude that perceptions of safety, road conditions, air quality, security, absence of provision for cyclists both on the roads and at place of employment, inconvenience and a residual stigma attached to cycling which associates it with childhood, scruffiness and lack of sophistication, are major factors (Unwin 1993).

Cycling is something of a poor relation amongst competing environmental policy themes. At a national level, the main engine of any policy on cycling is the Department of Transport (DTp), which lays down transport objectives, oversees legislation including the

Road Traffic Act 1991, New Roads and Streetworks Act 1991 and the Traffic Calming Act 1992 and attendant regulations. It advises local authorities on design matters, funds their transport schemes and holds responsibility for cycling within these activities.

The DTp has long been the target of opprobrium for groups interested in cycling. Traditionally, it has not viewed cycling as a mode of transport, merely as a recreational activity. As recently as 1991, in evidence to the House of Commons' Transport Committee, the DTp stated that 'it's not the Department's role to encourage people to cycle' (DTp, 1991). The National Road Traffic Forecasts for Great Britain made by the DTp did not include any mention of cycling (DTp 1989). In its National Travel Survey, the DTp relegated all journeys of less than one mile to a separate chapter, in the view of the Cyclists' Touring Club (CTC) marginalising one-third of all journeys and distorting the apparent contribution of different modes of transport (CTC 1992).

This apparent determination to relegate cycling to insignificance was matched, or perhaps caused by adherence to an orthodoxy with roots in the Thatcher era that saw spending money on roads as investment and on any other means of transport as subsidy. It was compounded by the DTp's own targets on accident reduction, which saw cycling as a high risk activity when compared to driving. Such an assertion was again challenged by interest groups (CTC 1992), who compared experience in the Netherlands where cycling accident rates are far lower, and correlate not with bicycle use but with car use.

Boulter (1994) argues that it was not until the 1980s, when concerns about congestion and the health problems associated with cars were joined by the environmental concerns of the post-Rio world, that the DTp's single-minded drive for ever greater volumes of road traffic was checked. The Sustainable Development Strategy for the UK (HMSO 1994a) includes the DTp in its scope, and although it contains little for the cyclist, it does nevertheless reflect a shift in the position of the DTp. The process began with its 'Killing Speed' programme and traffic calming initiatives. It gained a major boost in the annual TPP Circular to local authorities in 1993 which for the first time mentioned the need to reduce emissions of CO_2 and for the development of environmentally friendly forms of transport (DTp 1994). It also changed the way in which bids for funding could be made, away from a system which favoured investment in major roads at the expense of other transport systems. Some argue, as we have done with regard to environmental policy generally, that the shift originated not in a change of heart in the DTp, but in EU policy (Morphet 1994a), although it is also probable that the DTp was lobbied by local government towards the new system as a way of meeting local transport objectives more cheaply. The CTC point to the role of the Association of Metropolitan Authorities policy document, *Changing Gear* (AMA 1990), as playing a significant part in developing the new approach.

The TPP system effectively controls local authority transport capital expenditure, with each local highway authority being invited to submit a TPP bid to the DTp annually. Under the new TPP system, local authorities can submit packages of measures for funding, which can include infrastructure for cycling. A clearer pro-cycling signal came at the VeloCity Conference in Nottingham in 1993, where Roads and Traffic Minister Robert Key declared that he would be 'looking at local authorities' bids under the new "package approach" and I want to see a clear commitment to cycling'. Criteria by which proposals are judged at the DTp will, in the 1995/6 round, include easing of congestion and environmental impact and safety measures for vulnerable users. This should favour those packages which stress cycling provision (CTC 1993a).

Drift by the DTp away from its private car orthodoxy is evidenced most recently in Robert Key's June 1994 Cycling Statement, delivered at the launch of Green Transport Week (DTp 1994). The statement calls for action in a number of areas. In particular, local authorities are called upon to consider cycling in planning and traffic management from the outset. Mostly, however, the substance is largely a vague invocation to voluntary action on the part of employers, agencies and individuals.

In the late 1980s and early 1990s, the DOE became increasingly interested in the relationship between travel, land use and air quality issues. One of the results was PPG 13, jointly issued by the DOE and DTp, but originating in the DOE. It has a separate section devoted to cycling provision, with the admission that 'the level of cycling in the UK is significantly lower than that in a number of neighbouring countries, which have taken steps to make cycle use attractive as a day to day means of travel'. It calls for an 'effective network of cycle routes' and contains a number of suggestions as to how these might be achieved. Critics point to the fact that the guidance is strong on tackling traffic congestion, but weak on air-quality issues (CTC 1993a).

There is as yet little departmental integration on the issue of cycling. The Department of Health appears reluctant to encourage cycling for safety reasons, and seems unwilling to find a connection between public ill health and vehicle pollution whilst the Inland Revenue has declared a 5.6 pence per mile ceiling on cycle expenses, above which they are to be seen as income. The Department of Trade and Industry, despite its role in ACBE, seems as yet to be unaware of the environmental role of cycling.

A detailed survey of the TPP bids made by all local authorities has been carried out by the Cyclists' Public Affairs Group (CPAG 1994). It showed that the new rules for TPPs had produced a rise in the average proportion of bid devoted to provision for cycling, to an average of 0.76% per bid, three times the total in 1984/5. But local authorities surveyed still felt that the assessment criteria used by the DTp do not take into account the health and environmental benefits of cycling. It also revealed that many authorities, despite commitments to cycling, bid nothing for cycling provision. These included ten London boroughs, most of Merseyside and some County Councils. Some 20% of authorities were unable to say on what they would spend the money for which they had bid. Oxfordshire County Council made an assessment of its own bid, using different criteria, and demonstrated much greater financial benefit from the promotion of cycling.

Local authorities in the UK are relatively powerless to introduce radical approaches to transport when compared to some of their European counterparts. There are nevertheless many encouraging schemes being developed in a number of authorities. Amongst the Environment Cities, Leicester has made a particular theme of cycling provision. With the 1981 census revealing 3.8% of journeys to work being made by bicycle, the city has decided to target cycling as a way of reducing peak-time car commuting. The approach is to actively sell cycling to the commuter, and combine this with provision of new infrastructure, in collaboration with the Environment City private sector partners. Such infrastructure includes secure parking, showers and lockers at places of employment and a breakdown service for bicycles. Leicester is also working with the County Council to get an improved proportion of the next TPP bid to the DTp.

Boulter (1994) decries the current haphazard approach made to cycling provision, especially the reliance on cycle activists in devising provision: 'this approach . . . is indefensible as the main basis for planning work . . . no-one would assume that anyone who happened to drive a car was qualified to plan a road system'. He argues that planning for cycling is as complex as planning for public transport, and that planners must step into this breach with a professional approach.

Cycling shows some of the pitfalls in trying to superimpose environmental policies and programmes onto a system set up to meet quite different objectives, and in which 'environment' has been treated as an add-on. Comparing even the DTp's most recent utterances (DTp 1994) with the 'Bicycles First' statement of the Ministry of Transport, Public Works and Water Management of the Netherlands makes for depressing reading. As part of a 'Structured Scheme for Traffic and Transport', the Dutch statement explicitly links global environmental issues with quality of life; talks of the need to reduce mobility and promote accessibility; and makes specific, targeted commitments to increase cycling and the infrastructure for it.

Future directions for environmental policy

There is a real need for better communication and partnership between local and central government. The Central/Local Government Environment Forum (CLGEF) has existed since 1989; however, it has met only twice a year and according to Hams (1994) it fails in two main areas: only the DOE attends on behalf of government, and it does not integrate local authorities within government environmental initiatives in the business and voluntary sectors. Given the centrality of cross-sectoral partnerships to achieving sustainability, better consultative structures need to be built.

Looking further ahead, there may well be a change of government in 1996/7. Any future Labour government will look to the party's recent policy statement, *In Trust for Tomorrow* (Labour Party 1994). This document promotes an interventionist and integrated stance on environmental policy-making; it recognises the connection between environmental, economic and social issues, develops the concept of 'environmental rights' for citizens and suggests an overall approach to sustainability modelled on the Dutch concept of a national environmental policy plan. This would put a Labour government in step with international environmental agendas, a fact which can hardly escape policy-makers of any political hue.

The need to reintroduce the idea of a national plan for sustainability becomes all the more urgent given the growing momentum of the European integration process. The substantial portion of the Fifth Environmental Action Programme which is yet to be implemented, means that central government has little choice but to take a planned approach more in step with its European partner nations. The trend in the EU towards a growing role for regional government means that a new regional tier is increasingly likely in the UK. Given that this tier is likely to be dominated by 'local government thinking', a national government that has not put its environmental policies on a more integrated and sustainable footing will become increasingly and irrevocably marginalised by local authorities taking the 'Whitehall bypass' to Brussels. This route will become all the more significant if the prediction that the Committee of the Regions is to become the second chamber of the European Parliament is well founded.

8

British Land Use Planning and the European Union

TIM MARSHALL

In the mid 1990s there is a strong temptation to refer to 'environmental planning', as an emergent or desirable form of planningin Britain and Europe, rather than to 'land use planning'. In this chapter this temptation is resisted. This is because it would give an inappropriate impression of solidity and agreement about such a term in a cross-European context. 'Environmental planning' does exist in certain forms in some northern European countries (e.g. Netherlands, Germany) as a counterpart to physical or spatial planning. It refers variously to planning of landscape, of pollution control or of overall ecological processes. But in Britain even this degree of statutory and practical entrenchment does not exist. In this chapter, therefore, the prime emphasis will be on the impact of the European Union on British planning practice in the more traditional sense. A later section will touch on the relationship of emerging environmental planning ideas in Britain and European influences – but this is a secondary theme.

The chapter is intended as a survey and a discussion of the question of the Europeanisation of British planning. For details, reference will be made to more comprehensive treatments, above all that by Davies and Gosling (1994) for the Royal Town Planning Institute. The survey will take the form of a categorisation of, and

Environmental Planning and Sustainability. Edited by S. Buckingham-Hatfield and B. Evans.
© 1996 by John Wiley & Sons Ltd.

commentary on, the influences on British planning (as summarised in Figures 8.1 and 8.2). The discussion will run on a different, often more speculative level, seeking to go behind the immediate processes. This level will first be situated within broader ideas current in European economic and geographical thinking.

Space, environment and planning

Land use planning remains, in Europe in the late 1990s, an intensely contested and conflictual field. This is so in several, conceptually distinct but interrelated respects.

1. *Glocalisation*. This word, unpleasant though it may appear, is a useful term for catching the changing interplay of the global and the local (Swyngedouw 1992). It is based on the contention that capital has been moving faster and with more scope for choice,

Type of Influence	Initiative	Preparing Agency
Directly Spatial	• European Spatial Development Perspective	Committee for Spatial Development (and consultants)
	• Europe 2000+	DG XVI (and consultants)
	• Planning Systems Compendium	DG XVI (and consultants)
	• URBAN Community Initiative 1994–1999	DG XVI
	• Annual Sustainable Cities Reports	Urban Environment Group of Experts/DG XI
	• Sustainable Cities Best Practice Guide	Urban Environment Group of Experts/DG XI
	• Conferences and network co-ordination on sustainable cities	Urban Environment Group of Experts/DG XI
Not Directly Spatial	EU Sectoral Programmes, especially:	
	• Structural Funds	Mainly DG XVI
	• Trans-European Networks	Mainly DG VII & DG XVII
	• Environmental integration policies especially Fifth Environmental Action Programme (1992)	Mainly DGXI

Figure 8.1   EU initiatives with influence on British planning

Elements of British Planning	Degree of EU Influence
Systems	Very limited change, via introduction of environmental impact assessment
Policies	Change very variable, ranging from strong influence in more environmental fields (pollution, nature protection, coastal planning), to lesser effects on economic development and urban policies. But some growing influence in some regions and localities via cross-national spatial planning, and sustainable cities initiatives
Institutions	Limited influence on most parts of core planning institutions (ministries, local authorities etc), but growing weight of links to new EU based or funded initiatives (Committee for Spatial Development, Urban Environment Group of Experts, Committee of the Regions)
Discourses	Increasing effects of 'ecological modernisation' discourse (within and beyond EU initiatives), but countered by powerful economic development and competitiveness strands of EU corporatism and neo-liberalism
Actors	Significant influence on some NGOs and pressure groups, and certain sections of central departments and local authorities; profession beginning to make potentially significant responses

Figure 8.2   Degrees of EU influence on elements of British planning

since the 1970s. This leaves cities, regions and countries more exposed to rapid devaluation than before, and more determined to intensify efforts to improve the attractive qualities of places. Competitiveness of places reflects competitiveness of firms. This results in the entrepreneurial city (Harvey 1989) and the rise of the regional (Harvie 1994), but also in the intensification of inter-state competitiveness in certain respects, and a line of argument about competitiveness and industrial policy at EU levels (Nicolaides 1993).

Glocalisation, to the extent that it is emerging and exists, could have powerful effects on planning practice and the degree of Europeanisation of planning. This is because it changes the articulation of space and the importance of infrastructures, at several levels. It tends to emphasise 'local–territorial recon-figuration' (Swyngedouw 1992, p. 61) – the efforts by cities and regions to adapt themselves to the perceived needs of foreign investors, in whatever alliances appear most beneficial. The stronger glocalisation becomes, the more complex the Euro–state–region–locality articulation of land use planning practice is likely to be.

2. *Macro-spatial changes.* To a degree independently of glocalisation, Europe is being transformed by spatial shifts at the large, often continental scale. The restructuring of central and eastern Europe and the changing of the spatial economy of western Europe (by means of road, rail, tunnel and bridge building, and perhaps by telecommunications) are the visible signs of these shifts. Overall these changes are seen as giving the European continent greater mainland and maritime depth, and as boosting sea transport. There is little consensus on the impacts on Britain and Ireland, but their assessment has been affecting local and regional planning policies in many areas.

3. *Environmental policies.* At the start of the 1990s it appeared to many that a process of 'environmentalisation' of planning in Europe was likely to occur. This was based on emerging understandings of cross-frontier relationships, and on the formation of a single market space, with implications for an even 'environmental playing field'. The force of deregulatory and neo-liberal politics and the contradictions of economic changes since that time have been such as to hold back such shifts to a considerable extent. Nevertheless, environmental policy development has continued to diffuse at EU, state and local levels, in part under the banner of post-Rio (Agenda 21) initiatives. Here, therefore, pressures are globally as well as EU led.

4. *Deregulation.* The reduction of state intervention in many economic and social fields continues to affect the policies and instruments within land use planning, in Britain and elsewhere. The regulation of space, via planning systems and investment or other regulatory instruments, remains a relative bastion of public action. But its use, for economic or social or environmental ends, remains intensely contested. This is one reason why many of the EU initiatives described below have been, or are likely to be, so heavily fought over.

Together these four tendencies form a force-field within which the relationship 'planning and Europe' is developing. All are highly political and politicised. Any appearances to the contrary in the survey that follows are a testimony to the relative autonomy of particular discourses and policy fields, or to effective political and bureaucratic camouflage. The chapter's final section will reflect briefly on the degree of openness that these forces generate for the future of 'planning and Europe'.

Directly spatial EU influences on British planning

Here I will distinguish between the more directly and less directly spatial effects of current EU policies on British planning. The distinction is quite partial, but serves to divide those policy areas that are more specifically regional and urban from those within broader fields. Figure 8.1 presents the main types of influence. The distinction also serves to emphasise that the less directly spatial elements remain, probably by a long way, the most important ones for British planning. Nevertheless the treatment here will place relatively more stress on the emerging spatial/urban agendas. These can be divided into initiatives affecting the higher (regional) and lower (urban) levels. Discussion of specifically rural dimensions can be found in Davies and Gosling (1994).

It should be noted that, just as the Single European Act brought environmental policy directly into the EC's sphere of competence, the Act of European Union (Maastricht) has the following phrasing in Section 130(s): 'The Council, acting unanimously on a proposal from the Commission and after consulting the European Parliament and the Economic and Social Committee, shall adopt . . . measures concerning town and country planning, land use with the exception of waste management and measures of a general nature, and management of water resources' (Commission of the European Communities 1992c). This section appears within Title XVI, Environment. Whilst town planning's arrival is clearly hedged around with exceptions and the requirement, normally, for unanimity, it is at least present for the first time.

This does not mean that one would be likely to find much consensus on what *should* be the role of Euro-level planning. Williams (1990, p. 16) argued that the 'vast majority of planning policies and decisions can operate totally satisfactorily within existing national, regional or local jurisdictions'. But he did identify three roles for European promoted planning. Two were relatively limited, for most purposes. These were for trans-frontier planning and for the exchange of information; these would not affect sovereignty, but would require better inter-state communication. There was also though, he felt, a 'policy level', for matters which were 'strategic' for Europe as a whole; he suggested these could relate to policies for large-scale infrastructures (airports, seaports) or for the reduction of regional disparities, or for nuclear waste or nuclear power stations. The difficulty with these criteria is that they could justify very wide planning intervention for economic, social or environmental purposes, and so undermine Williams's conclusion. Since that time the balance may have tipped slightly, in some countries, towards a broader interpretation of the scope for European planning strategies, as the following sections will show.

European spatial development

The passing of the Single European Act and the prospect of '1992' led to an intensification of work in the European Commission on regional prospects (Martin 1990; Williams 1991, 1993; Davies and Gosling 1994). In 1989 a new regional/spatial planning unit was set up in DG XVI. Under Article XI of the European Regional Development Fund Regulation 4254/88, a number of Community Initiatives were funded, leading to the publication of *Europe 2000* (Commission of the European Communities 1991). This was essentially an analysis of trends within the EC, rather than a presentation of policies. But it was accompanied by more policy-orientated exercises such as the document *Perspectives in Europe* (Rijksplanologische Dienst 1991), promoted during the Dutch presidency. *Europe 2000* had widespread publicity and led to two further initiatives. The first set up a Committee on Spatial Development in November 1991. This was made up of central government officials and Commission representatives and has since overseen the production of eight trans-regional and three external impact studies, and a compendium of member state planning systems, due to be published in late 1995. Eight new research studies were commissioned in 1993, on issues as diverse as internationally mobile investment, national transfer and financial equalisation mechanisms, and transnational dimensions of environmental protection. The second initiative, set up at the Council of Ministers with responsibilities for regional policies in May 1992, broadened the role of the Committee on Spatial Development to include the preparation of a European Spatial Development Perspective – a vision of territorial development in Europe as a whole, intended in part to guide planning of trans-European networks. In parallel the Commission proposed to produce Europe 2000+ in 1995, providing a more strategically orientated basis for influencing EU policies across all fields. These built on earlier efforts. Williams (1990) chronicles the attempts throughout the 1980s by the Council of Europe – the Torremolinos Charter (Council of Europe 1984) and the European Regional Planning Strategy (Council of Europe 1988/1992). A document produced by the German Federal Ministry for Regional Planning, Building and Urban Development (1993) showed how enthusiastic certain states were about these initiatives. The report was entitled *Spatial Planning Policies in a European Context* and was intended as a draft version of a final document to be progressed during the German presidency of late 1994. It recommended the formalisation of both the 'Physical Planning & Regional Policy' Council of Ministers and the Committee for Spatial Development. It pressed for regular reports on 'the territory of Europe', after Europe 2000+, and strong support for the rapid development of a modern settlement structure in east and central Europe. Within the EU, 'action areas for spatial planning policy' (p. 5) should be designated. The document in effect extends German Federal and Land planning concepts to the European level, noting that the co-operative creation of a well-targeted European spatial development policy would be made easier 'when all the Member States establish spatial planning goals for the long-term development of their territories' (p. 2). This would clearly not fit with British planning traditions (or government policy), and the German approach might well not fit easily with French or other traditions. But the German policy drive was clearly carrying on from where the Dutch had left the issues at the start of the 1990s. The Belgian Presidency had concluded after the Liège informal Regional Council in November 1993 that 'whilst respecting the principle of subsidiarity, significant progress could and should be made on a concerted planning strategy for the European territory' (EU Presidency 1993, p. 4).

Between 1988 and 1994, therefore, a whole new policy field was developed within the EU, despite the shifting sands of European politics during these years, and despite the fact that the authorising Councils were informal and the Committee on Spatial Development had an uncertain legitimacy. The implications for British planning of these initiatives remained unclear. The three trans-frontier studies affecting the UK (central capitals – SE England, the North Sea regions and the Atlantic Arc) all involved British planners at central and local levels. In the case of the Atlantic Arc, earlier networks linking the regions laid the groundwork for a separate programme, ATLANTIS, from 1993. However, British central government remained relatively unenthusiastic about these exercises and there is little sign so far that the initiatives, of themselves, are influencing British policies at national level. Each of the trans-frontier studies has trend and policy scenarios, presented on map bases; but any attempt to incorporate decisions based on these is likely to prove highly controversial in Britain. The emerging outputs of regional guidance may change this judgement in due course. Much more likely to be influenced are the regional groupings of local authorities, perhaps particularly in the South East, the South West, Yorkshire and Humberside, as they pursue their own economic development strategies. But here aspirations often reach further than real possibilities, without central government leadership.

The emergence of the Europe 2000+ and European Spatial Development Perspective documents, and the formalisation of the regional ministers Council and the Committee for Spatial Development, would lay the groundwork for a new round of policy-making. But the outcomes of both of these remain uncertain at the time of writing.

More long-term initiatives on the planning systems compendium, the European research institutes network and the proposal for a European Planning Academy may all be influential, eventually. Behind the first is primarily an attempt to inform planners and developers of systems across nearby borders – something of more immediate interest to say Dutch planners than to those in Britain. But pressures for some degree of harmonisation of planning systems could possibly follow on, as Williams (1991) discussed. The research institutes network and a European Planning Academy could boost the so far gradual and uncertain Europeanisation of planning education and research, by building up a more coherent policy community. At present, many links are partial and bilateral (such as British links with central and eastern Europe), but this initiative, like others, reflects the EU's recent drive to meet 'subsidiarity' demands by promoting networks, rather than depending only on central governments. This aspect clearly fits with both the glocalisation and deregulation tendencies outlined above, but whether 'network power' can really challenge the powers of central governments remains unclear.

Urban development and the environment

The European Commission has led two somewhat separate streams of policy-making for cities.[1] One has been on urban environments, flowing from DG XI's production of the Green Paper on the Urban Environment (CEC 1990). Extensive debates, by no means all favourable to the Green Paper's promotion of 'compact' cities, were followed by the formation of the Urban Environment Group of Experts in October 1991. This had 12 central government representatives, as well as 12 members from interest groups and some enthusiastic cities (Davies and Gosling 1994; Fudge 1992, 1993). After the decision not to produce a White Paper on the subject, the expert group eventually picked up speed, leading to a conference in Aalborg in May 1994 and the linking of existing urban networks dealing with environmental issues (EUROCITIES, ICLEI – International Council for Local Environmental Initiatives). The strategy was therefore to build on Agenda 21 and similar local initiatives in order to develop a dynamic for sharing information and eventually, no doubt, creating direct EU programmes. In 1994 it seemed that the Commission had found a constituency, and in effect a lobby, with considerable cross-EU resonance, in which British planners were taking leading roles. Again though, it was unclear where this would lead in terms of concrete EU and state policies and funding. For the present, work such as that of the

---

[1] I am grateful to Ian Clark of DG XI, to Michael Cronin of the Department of the Environment and to Colin Fudge of Bath City Council and the Urban Environment Group of Experts, for help on the urban initiatives.

sustainable cities group hardly figures on the agenda of the Environment Ministers Council meetings. But the potential influence
on the ways of thinking about and discussing urban issues is
considerable.

This would be especially the case if DG XI's initiative is fully linked
to DG XVI's ongoing involvement with urban projects. This began
in 1989, under Article X funding, and covered 30 cities, primarily in
'mainline' economic and social development fields (Davies and
Gosling 1994, pp. 54–55). These pilot projects were followed by the
URBAN initiative (CEC 1994b), taking 600 MECUs of structural
funds. It appears that the two DGs are driving in partially different
directions, one likely to be administered by those on the economic
development side of planning work, the other by more environmentalist planners. Any dissonance may be reduced by the proposal for DG XVI to prepare an integrating Commission document
on urban issues, bringing together the interests of a range of DGs
(energy, telecommunications, social/poverty, transport, as well as
XI and XVI).

There is little sign that 'sustainable cities' as a slogan has effectively
united the contradictions inherent in the real operations of cities.
But the urban networking movement (Camhis and Fox 1992;
Dawson 1992; Marlow 1992; Davies and Gosling 1994, pp. 55–58)
may be gradually developing some ability to construct new arguments and political formulas. The problem it confronts lies in the
parallel pressures for intense urban, entrepreneurial competition,
tending to lower social and environmental standards within a
frontier-free market-place. Cities, even compared with regions,
normally have relatively weak tax bases, which they are afraid of
expanding. 'Sustainability planners' working with the EU are
therefore on exciting, but difficult, ground.

Non-spatial EU influences on British planning

It is increasingly clear that all EU programme areas have impacts on
British planning, from competition policy to telecommunications,
social policy to fisheries. It could be argued for example that the
processes of internationalisation in finances and services, partly due
to the Single European Act, have been responsible for a certain

degree of Europeanisation of property markets (Wood and Williams 1992). This then may push towards some harmonisation of planning regulations, if developers and their agents are active in several countries and press for reduction of such 'barriers to free movement'. In that case DG II (for the Single Market) would appear as the key part of the EC Commission, for land use planning.

Davies and Gosling (1994) deal extensively with the most important programme areas, especially with the Structural Funds – the part most familiar to British planners, and increasingly so as more parts of the UK have recently become eligible for assistance. In many European regions transport and energy networks have become centrally important elements of EU planning and funding. The UK's lesser physical connections and lesser political enthusiasm (at central level) have no doubt reduced the impact of the 'Trans European Networks' programme on the UK. But debates on transport and energy, and funding, especially for pilot projects, have made these areas of EU policy by no means insignificant, as the RTPI report demonstrates.

In part this impact has been mediated through EU environmental initiatives, and I will concentrate on these here. This is not to imply that they are the most important in concrete terms; on the contrary, the other programme areas must have a much greater weight in total. But the EU's environmental push has in recent years aroused greater enthusiasm and interest from British planners and arguably has radical implications for land use planning.

It has been suggested that British environmental policy has been extensively Europeanised, by policies emanating from the EC since British accession (Haigh 1989). The cross-over from these more precisely environmental policies to land use planning policies is both apparently strong, and very complex. Whilst it is clear that British law and practice on air and water pollution, and nature protection, has been transformed by EC influences, the same cannot be said of land pollution or spatial regulation. Only in one, well known, respect has the land use planning system been changed significantly by EC membership, that of environmental impact assessment (chronicled by Williams 1988). Even here though the new instrument has been stapled onto the existing system, with uncertain and variable results.

In the field of policies the effect has arguably been greater. First the gradual EC pressure through the 1980s was one factor in producing the first relatively comprehensive policy document, the 1990 White Paper (DOE 1990). This then began to have effects on the whole range of British national policy instruments, especially the 'new model' Planning Policy Guidance Notes of 1992–1994. These had in turn been no doubt affected by the parallel efforts emerging from the EC Commission, especially the Green Paper on the Urban Environment (CEC 1990) and the Fifth Environmental Programme (CEC 1992a). This programme calls for 'an appropriately comprehensive planning/development/environmental protection framework' to achieve 'the socio-economic development and ecological health of a country, region or locality'. This would involve 'an identifiable sequence starting with national and regional economic plans and ending with local physical development and environmental protection plans'. This page of the programme (p. 66) includes one of the more positive endorsements of planning's role to be seen in official documents in recent years: 'In the endeavour to achieve sustainability, the planning functions and the public authorities in whom responsibility is vested must ensure an optimisation of the "mix" of industry, energy, transport, human habitation, leisure and tourism, ancillary services and supporting infrastructure which is consistent with the carrying capacity of the environment'. Whilst in Britain government policy would not allow exploration of the possibilities hinted at in this sentence, it does serve as a tantalising glimpse of what certain parts of an environmental plan might contain. This no doubt explains why, along with Agenda 21, it has therefore been a key document in developing broader thinking on environmental planning in local government – as shown by its widespread distribution by bodies like the Local Government International Bureau.

The new thinking in British planning, whether in general approaches such as that of Blowers (1993), or in attempts to produce methodologies (Council for the Protection of Rural England 1993) all draw to some degree on European sources such as the Fifth Programme. But this dependence should not be exaggerated. The methods advocated in the above CPRE document, for example, which are meant to advance 'capacity planning' as a new tool, draw as much on 'indigenous' planning traditions (however camouflaged) or on academic work on the boundaries of economy and ecology, as on EU initiatives (one clear source being Jacobs 1991). The worlds of planners in NGOs and in local authorities in Britain intersect in complex ways with global and local environmental movements, such that these do not match at all exactly the geographical or political reaches of the EU.

If there is a new 'environmental planning' in Britain, it will therefore most likely be to a significant extent a home-grown variety, emerging from the clashes and efforts of many central and local governmental, and non-governmental actors. It would therefore be going too far to say that British planning is being Europeanised by EU environmental policy, even though the effects of the latter have been significant. It is expected that the EU's Fifth Programme (meant to run 1993–1997) will be examined for possible revision in 1995, when the initial attempts to integrate its objectives into the full range of EU policy areas will have been completed. It will then be timely to consider whether land use or spatial planning dimensions will be given a higher, or lower, profile. This will depend in part on the evaluation of the effects of environmental impact assessment, and what scope is envisaged for strategic impact assessment, whether of EU or national state programmes.

Extent of Europeanisation of British planning

The current degree of Europeanisation can be examined from another angle, by focusing on five dimensions (Figure 8.2). In this way the very uneven impact of the EU becomes clear.

Systems

There has been some discussion of the possibility that European planning systems might have been converging in recent years in certain respects (Davies *et al.* 1989; Keyes *et al.* 1993; Davies and Gosling 1994). In particular it has been thought that the balance of flexibility/rigidity in development plans and control might have been moving to a point mid way between the British and the more rigid continental types. Other sorts of convergence might be detected in practices on planning gain. However, any such shifts are likely to be theorised as resulting from changes in the relative powers of developers and public agencies, or possibly from internationalisation of property development. The EC has had no direct effects on planning systems, beyond the environmental impact assessment procedures introduced in the 1980s.

Policies

British land use planning policies are made primarily within a nationally framed planning agenda, by central and local governments. For example, a shift on a core development issue such as out of town shopping centres would, till now, have emerged essentially from a British political debate or argument. But immediately, in this case, the indirect, often slow moving, effects of EU environmental policies are visible, mediated through debates partially sparked by EU initiatives, such as the Green Paper of 1990 or the Fifth Environmental Programme of 1992. Sometimes these initiatives have been carried into the British policy community by NGOs, sometimes by local authorities, much less commonly by industry, developers or central government.

In the absence of directed research, it is hard to say which British planning policy areas have been more Europeanised – other than that those 'nearest' to major environmental concerns are the most likely candidates. These would include the relationship to pollution control, coastal policies and wildlife/nature protection. Meanwhile the Structural Funds' impact on regional and local economic development policies will remain significant.

The 'emerging agendas' in the spatial and urban policy fields, discussed above, remain an uncertain contribution to British planning. In some regions (Kent, the South West) the European spatial development perspective and Europe 2000+ will be viewed with tremendous interest, just as some cities see themselves in the vanguard of the 'sustainable cities' movement, partially promoted by the EU Commission. In other regions and localities these initiatives will remain invisible or secondary, and under the current government neither initiative is likely to be given great support centrally or incorporated into national policy instruments.

In part these limits to the Europeanisation of planning policies can be traced to the UK's island position. This does not deny the long-appreciated facets of peripheral position, or the importance of the Channel Tunnel and air and sea links, or the significance of long-range and global pollution. It simply points to the likelihood that the trans-frontier planning processes will remain of lesser importance than in many mainland states.

More generally the neo-functionalist argument that social and environmental EU policies tend to follow from the economic (especially single market) initiatives appears at least uncertain. If a brand of neo-liberalism blocking any such tendency is politically dominant, then these policies, and land use planning policies associated with them, may not be developed in the near future to any significant extent.

Institutions

The main institutional bases of British planning remain essentially nationally orientated. This is so in the Department of Environment where the International Unit in the Planning Directorate consists of two civil servants – even though many others work at times on EU issues. It is also so in local authorities, and in most planning consultancies, even though increasing numbers of authorities have created European units, initially at least mainly focused on securing EU funding. However, it can be argued that the institutional bases in or near the EU which relate to land use planning are beginning to reach a critical mass. The informal Council meetings of regional ministers (since 1989), the Committee on Spatial Development and the Urban Environment Group of Experts (both since 1991) and the Committee of the Regions (since 1994) all add some weight to the already existing processes of contact, inside EU fields (European Parliament Committees, the Economic and Social Committee, EU funded city and region networks or associations) or outside (European Council of Town Planners, the Association of European Schools of Planning, Council of Europe).

The bases may become more formalised and significant for land use planning, as many member states and planning professionals appear to wish, or they may remain at the relatively embryonic stage reached in 1994. The institutional review of the EU proposed by 1996, which could affect in particular the status of the Committee of the Regions, will be important in this respect. From this point of view the accession of the Denmark, Sweden and Austria is likely to push towards some greater impetus on environmental and cross-frontier co-operation, though not necessarily on regionalisation.

Discourses

The EU is full of competing and contradictory discourses, some deregulatory and neo-liberal, others more social democratic and corporatist. The prospect for land use planning's Europeanisation is partly related to the shifts and struggles within these discourses. One emergent framing of economic and environmental issues, that of 'ecological modernisation', is of special potential importance for planning. But, as Hajer (1994) has discussed, this discourse contains various conflicting formulas, some more technocratic, some more connected to 'base democracy', some in reality concerned primarily with continued production growth and 'sustained mobility'. The emergent 'sustainable cities' agenda has proved one effective framing of planning issues, but it is building on the ambiguities of Brundtland's sustainable development, and ecological modernisation's uncertain direction.

A review of significant recent pronouncements by British environmental ministers, local politicians and professional planners would no doubt reveal the influence of the ecological modernisation discourse, in uneasy tandem with the powerful framing of 'competitiveness', at all levels. Here, therefore, the EU's contribution is again contradictory, as it is quite as much involved in the discourse of competition as that of environmental moderation. In the case of Britain, though, one can say that its influence has been, in discourse terms, greater on the ecological side than the economic. In concrete (Structural Funds) terms, the opposite will have been the case.

Actors

It is evident that the degrees of EU influence on different actors vary greatly. This is so across and within levels of government and between interest groups. Broadly speaking NGOs and local authorities have been more strongly affected in their outlooks and to some extent practices than central government. An examination of the annual reports after the 1990 Environment White Paper, or of the UK Sustainability Strategy (HMSO 1994a) shows that land use planning issues were retained mainly within a UK rather than an EU frame.

Nevertheless, parts of central government, such as the Environmental Protection Directorate of the Department of Environment, or the Department of Trade and Industry sections dealing with Structural Funds, are quite fully integrated into EU policy-making processes. It is notable though that the above directorates' links to the core planning areas of the DOE are relatively weak, such that each directorate tends to deal with the EU via relatively separated channels. This partly reflects the fact that environmental issues only came to play an important role in the Planning Directorate some years after they were fully established in the Environmental Protection Directorate.

The planning profession itself reflects this unevenness, with a growing band of Euro-enthusiasts balanced by the main weight of the profession still tied to more precisely UK concerns. The publication of the 1994 RTPI report by Davies and Gosling may have some effect on this balance, backed by the regionalist and sustainable cities movements. The RTPI report advocated a proactive stance towards Europe, based on an analysis of the increasing significance of the EU's impacts. It recommended:

1. increasing awareness of EU developments, and adequate responses within organisations such as the European Council of Town Planners;
2. supporting exchanges of experience and practice across Europe, including consultancy opportunities;
3. ensuring the British profession is equipped by its education and membership policies to work effectively in the new Europe.

So far the new breed of planning professional is hardly to be seen, even if there is greatly increased awareness of EU issues, especially amongst newly qualified planners. It can be expected that the great national variation in systems, policies and discourses (including languages and cultures) will hold back the speed of change, however determined the RTPI's response.

Summary and prospects

Some commentators consider that Europe has become a core dimension of British planning. Williams, for example, has argued that 'the ability to think European, to conceptualise Europe as one planning subject is essential' and that 'the EC is in many senses one jurisdiction' (1994, p. 359). He suggested that there was now 'policy development backed by legislative, executive and judicial arms of government operating at the EC level' (1994, p. 348). Morphet argues, on the more specifically regionalist dimension, that 'the regional dimension is now one which is with us in the UK' and that 'whatever system is developed in the UK, it will need to be Eurocentric, otherwise resources may not follow' (1994, p. 19).

I would argue that developments up to the present suggest a more sober conclusion: that environmentalism, structural funding and trans-European networks have been affecting the policies and outlooks of British planning, carried through a gradual institutional development, but that the advance remains highly uneven. There is some degree of lack of synchronisation between systems, policies and outlooks or aspirations. But the essence of the British planning system and practice remains in a nationally guided policy framework.

So far there has been little firm development of a new form of 'environmental planning' in some wider senses, despite the often heroic efforts of civil servants writing annual reviews and local planners revising structure and local plans. Such a new form of planning would need a social and economic refocusing as much as an ecological one, and that is incompatible with the main current drives at British government and EU levels, whatever the efforts of DOE or DG XI.

However, these conclusions are somewhat static, and to get a grasp of the really greater openness of the future, it will help to return to consider the four dimensions considered at the chapter's beginning. These all related to the shifting economic forces in Europe, the changing geographical balances and the powers and wills of states or sub-state political bodies to affect these economies and geographies. A key political dimension in this dynamic in the 1990s has been the force of environmentalism, in all its dimensions (as social movement, industrial response, governmental discourse).

Shifts in land use planning practices would be quite different, depending on the variations on these four dimensions. For example, a more intense process of 'glocalisation' (in the sense of heightened globalising pressures and local proactivity) would have quite different implications for planning if accompanied by heavy state deregulation, and/or by a strong environmentalist drive. A slowing down of the construction of major communications systems could hold back 'glocalisation' and could generate less intense patterns of spatial concentration or pressures to peripheralisation. Land use planning, in Britain as elsewhere, is caught up in this parallelogram of forces, and *one* factor mediating these shifts is the EU, as political process.

On a more fundamental level, however, the EU's future will depend on the development (or absence) of economic and political crises across Europe (west, central and east), crises which at the same time must inevitably now have global dimensions. Whilst in very general terms one may expect that European capital will find some new 'spatial fixes' (in Harvey's (1982) sense), in eastern and central Europe, this will not necessarily circumvent powerful crises. These could change the direction of EU policy, which has been, in the 1990s, predominantly neo-liberal but with secondary social and environmental facets. The forces behind glocalisation and deregulation could therefore change in such crises. In the past, town planning systems in western Europe have been created and reformed primarily in the face of fundamental social and political crises – first those caused by industrialisation and urbanisation, then those accompanying the demands for radical state intervention following the two world wars. It is quite possible that some equally fundamental challenges will emerge in the wake of the dramatic changes of the 1980s and especially of 1989. These challenges would then condition the Europeanisation (or otherwise) of land use planning in ways only loosely based on the processes described in this chapter. In the meantime planners would do well to keep close track of EU developments, especially including the spatial and urban dimensions emphasised here. However, this should not be to the detriment of pushing new specifically 'local' forms of environmental planning.

9

The Constraints on Sustainability Planning in the UK

DUNCAN McLAREN

The key components of the sustainable development policy framework which are necessary at all levels of government, and in all agencies contributing to sustainable development are as follows:

- accountability and transparency accompanied by freedom of information and other statutory rights to enable participation;
- co-ordination of policy (including effective environmental assessment), integrating environmental and economic goals; and
- timetabled target setting, reflecting environmental capacity, supported by a regulatory framework within which a package of measures (including demand management) can be used to meet and monitor those targets.

This last component can be summarised as *sustainability planning*. Having reviewed the extent to which these components have been achieved in the UK, this chapter will assess the potential and implications of implementing sustainability planning, and the constraints operating upon it.

Sustainability planning

The principles of sustainability planning are (at least in theory) simple (Gee *et al.* 1990; Jacobs 1991). Sustainable development can

Environmental Planning and Sustainability. Edited by S. Buckingham-Hatfield and B. Evans.
© 1996 by John Wiley & Sons Ltd.

only be achieved if human activity is kept within the constraints set by environmental capacity. If technical information is poor or lacking, then to locate those constraints, the precautionary principle must be applied. From such sustainability constraints, political planning processes are needed to set targets which can be met through the application of a range of appropriate policy tools.

Environmental capacity

Conventional concepts of economic growth and development based on unlimited resources have been challenged by a growing understanding that one fundamental resource – the environmental capacity of the planet – is, in fact, limited.

Environmental capacity is the ability and adaptability of the environment to provide the physical and non-physical resources humans need, such as the provision of energy and raw materials, the absorption of wastes, genetic diversity and fundamental life-support services such as climatic regulation.

Because the current allocation of environmental resources does not take the interests of future generations or developing countries fully into account, we need to plan the use of these resources. In particular we need to put into practice the idea of futurity, which demands that we take into account the interests of future generations who cannot be consulted about their needs or the values they would place on environmental resources. This clearly distinguishes sustainable development from previous environmental initiatives in Northern countries. In practical terms, it means that the beauty, productivity and resilience of the environment is maintained in trust as minimum 'capital stock' of environmental capacity for future generations.

Concern for the impacts of economic development on the environment has resulted in attempts to 'balance' economic growth and environmental protection. But a process in which environmental capacity can be traded off against growth or development cannot guarantee sustainability. This is the false, or as Pearce *et al.* (1989) and others describe it, 'weak' version of sustainable development.

The true version stipulates that these environmental consider-ations should be seen as a constraint on other goals: meeting en-vironmental goals is an essential pre-condition of meeting other goals. In other words, these conditions are sustainability con-straints. The components of environmental resources that lie beyond the constraints should therefore be inviolable. Such com-ponents (for example, the minimum area of natural habitat needed to protect biodiversity or the minimum level of stratospheric ozone needed to protect health) can be described as critical natural capi-tal, which must be maintained.

Sustainability constraints

The limited nature of environmental capacity sets constraints to our use of environmental resources. These limits are necessary to protect the ability of the environmental resources to meet the func-tions we rely upon: 'for each element of environmental capacity a maximum [or minimum] stock or flow level can be identified, beyond which environmental capacity begins to decline. For an economy aiming at sustainability, these maxima effectively be-come constraints' (Jacobs 1991, p. 101).

These are sustainability constraints. If they are breached, then the function of that resource will be lost, whether slowly or rapidly, temporarily or permanently. The depletion of stratospheric ozone by chlorine compounds and the threat of climate change are good examples of the implications of such constraints: going beyond the environmental limits threatens catastrophic change.

Although we do not always have the understanding and informa-tion necessary to identify the level of these constraints accurately, in many cases we can identify practical limits. A methodology can be developed independent of definition of the exact targets, sup-ported by the precautionary principle. Once such limits are lo-cated, whether by science or policy, then targets can be set to avoid breaching these limits.

To reduce our dependence on the precautionary principle we need improved measures of environmental capacity. The study of crit-ical loads for aquatic systems provides one example of how more accurate measures can be developed (e.g. AWRG 1989; Edwards *et al.* 1989). Rates of resource use within environmental capacity can be determined for renewable and non-renewable resources (Jacobs 1991). The capacity of the atmosphere to absorb greenhouse gases has been heavily researched and significant progress made (e.g. Krause *et al.* 1989; Houghton *et al.* 1990; SWCC, 1990). To develop equally practical measures for aspects of partial or less confined systems is a greater challenge.

The issue of biodiversity conservation illustrates the blend of political and scientific methods that is needed to determine constraints. Biodiversity forms a critical part of environmental capacity. Not only is it fundamental to ecosystem functioning, but it is critical for the provision of new genetic material for agriculture and forestry, and for a range of foodstuffs and medical products (Rice 1993).

A continuing decline in biodiversity in the UK is indicated by the 1990 Countryside Survey (Barr *et al.* 1993), 'local and detailed surveys [which] reveal losses of the rarer and conservationally important habitats' (HMG 1994a, para. 13.12) and continued damage to the UK's network of SSSIs. The statutory nature conservation agencies reported that 319 SSSIs were damaged in the year 1992/3 (English Nature 1993; Scottish Natural Heritage 1993; Countryside Council for Wales 1993). These figures include all damage categories: total and partial loss, long-term damage and short-term damage. Government statistics often exclude the last category (defined as damage from which the site may recover within 15 years). In fact, the figure of 319 sites is an underestimate because the monitoring was inconsistent and incomplete. In Scotland, damage from neglect or by overgrazing was not recorded, although it was the most common type of damage recorded in England and Wales. In England, less than half of the total number of sites were monitored. Damage from the effects of acid rain was not monitored, even though this is one of the main factors affecting biodiversity in the UK (HMG 1994a, para. 13.21).

The environmental capacity of biodiversity can be translated into a sustainability constraint through the mechanism of habitat designation and protection. In the UK, SSSIs can be seen as the critical natural capital. The Nature Conservancy Council (NCC 1989) define them as the minimum area necessary to maintain biodiversity in the UK. However, government policy disputes this interpretation. The measures used to protect SSSIs provide for a process in which economic benefits from damage can be used to justify it. The nature conservation agencies have been forced to enter into expensive management agreements to obtain appropriate management of sites or in many cases simply to prevent damaging operations (Rowell 1991).

This is typical of UK policy. Rather than identify inviolable limits to polluting activities, the Sustainable Development Strategy refers to balances of costs and benefits (HMG 1994a). However, no clear indication is given of how the government intends to decide that the benefits outweigh the environmental costs or that a particular 'feature' of the environment needs to be treated as inviolable. No framework is offered for arriving at such judgements, particularly where economic benefits are received by rich citizens now and environmental and social costs fall in the future.

Development and equity

For much of the world's population, sustainable development is about current (rather than future) access to environmental and other resources. The developed countries control and consume the vast majority of those resources. Once it is recognised that they are limited, redistribution becomes a critical issue. Indeed it was perhaps the most critical issue at Rio. Without access to such resources, development becomes impossible.

Development is not the same as economic growth. Development involves an advance in utility and well-being (which may include monetary income), preservation or advance in freedoms, and increasing self-respect and self-esteem (Pearce *et al.* 1989). Increasing equity is also necessary: 'The well-being of the most disadvantaged in society must also be given greater "weight" in a developing society: if average well-being advances at the cost of a worsening of the position of the most disadvantaged it seems reasonable to say that such a society is not developing' (Pearce *et al.* 1989, p. 29).

But environmental equity means more than this. The identification of 'sustainability constraints' can be interpreted through the concept of 'environmental space'. This notional space is composed of all the environmental resources on which we can draw. Sustainability constraints limit the size of that space.

We can calculate how much of that space the average individual takes up, or consumes – and thus what a fair share of environmental space might be (FOE Europe 1994). Where our activities have already placed us at the boundaries of environmental space, any increase in consumption by the worst-off must be compensated for by a reduction elsewhere. To permit development in the developing countries it is generally the case that a reduction in Western consumption levels is required, either directly, or through efficiency gains.

Efficiency and sufficiency

There are two broad categories of measures designed to bring human consumption of environmental space within sustainability constraints: using environmental space more efficiently, and simply consuming less of it. These strategies can be termed 'efficiency' and 'sufficiency' (Buitenkamp *et al.* 1993). Efficiency can be seen as meeting needs and aspirations through providing equivalent goods and services but with less consumption of environmental resources. Sufficiency implies consuming less goods and services, but may involve meeting needs and aspirations in different ways, or obtaining some compensatory improvement in welfare or quality of life.

The trade-off between efficiency of use and total use is common to virtually all environmental resources. For example, in the transport sector emissions reduction targets can be translated into targets for increased fuel efficiency and for reductions in traffic volumes (Holman 1991). Reductions in car use may actually be compensated for by improved health and thus result in no overall decline in well-being.

Participation

At the other extreme of scale there are constraints to individual action. Participation is seen as a fundamental component of development in that people must be able to share in decision-making about the goals and means of development, and also be able to take an active role in pursuing them. Participatory democracy is funda-mental to our understanding of sustainable development as an ongoing process. Participation is central to Agenda 21: 'One of the fundamental prerequisites for the achievement of sustainable de-velopment is broad public participation in decision making . . . . Individuals, groups and organisations should have access to infor-mation relevant to environment and development held by national authorities' (UNCED 1992, paras 23.1–23.2).

However, the UK government has shown scant regard for particip-ation in decision-making. This is highlighted by the fact that the Sustainable Forestry Programme (launched alongside the Sustain-able Development Strategy) was compiled in secret with no public input. While broad consultation was undertaken on the Sustain-able Development Strategy, even where consultees showed over-whelming support for a particular course of action, their opinions on issues such as road-building, deregulation and SSSI protection were not accurately reflected in the Strategy (FOE 1994a).

More generally, the effectiveness of public consultation procedures in the UK would be disputed by many, such as the individuals and groups who have (with some notable exceptions) failed to per-suade the Department of Transport to drop proposed road-building schemes all across the country. New consultation mecha-nisms have been introduced as part of the Sustainable Develop-ment Strategy, in the form of 'round-tables'. But these have no guaranteed role in decision-making, and seem set to have less influence in the UK than in those countries, such as Canada, where the political system is more consensual.

However, public participation is fundamental to the changes in both life-style and attitudes that are widely seen as necessary to achieve sustainability (e.g. HMG 1994a). The policy framework is critical if such changes are to be enabled. A voluntary approach by government, reliant principally upon exhortation, will not only fail to bring about the desired changes, but risks stifling participation through the creation of a sense of disempowerment. Opinion sur-veys by Friends of the Earth and others have shown repeatedly that public willingness to take environmental action is high, but thwarted by the failure of government to act to remove the blockages to such action.

For example, without an officially validated labelling scheme for timber, shoppers will not be able to avoid buying tropical timber products which contribute to rainforest destruction – despite the fact that 58% of people would not buy a timber product if they knew it came from the rainforests and 83% of shoppers believe that shops should not sell such products (Friends of the Earth, 1992b). Similarly, without action to improve public transport, the 39% of all drivers who said they would use their cars less if public transport were improved will continue to drive (Lex Service 1994). Economic instruments are not an adequate substitute for regulation in encouraging participation. Over-reliance on economic instruments limits individuals' ability to take action as citizens rather than as consumers. Such citizen action is critical where environmental goods are in fact common property resources.

There are key tools available to facilitate participation: freedom of information, transparency of decision-making procedures, accountability or liability of decision-makers and an enforceable regulatory framework. If information is available in an accessible form to the public, reasons for decisions are clear to the public, and there are mechanisms whereby redress is available for the public, then the public will be empowered to participate in the process of development. The provision of statutory environmental rights is a fundamental requirement for effective participation but, as is shown below, the government has yet to introduce even an adequate Right to Know.

Freedom of information

The UK Sustainable Development Strategy considers neither the use of Freedom of Information as a policy tool nor the importance of transparency and accountability where decisions are made that affect our environments. Instead, it merely notes that the public often want better information. In the past the government has gone so far as to suggest that citizens have a right of access to the information about the environment they need in order to act as an environmental watch-dog: 'That information is the citizen's right and the Active Citizen will use that right constructively' (Major 1991).

While the UK has introduced local authority legislation and regulations to implement the European Directive on Freedom of Information, obliging public sector bodies to permit access to environmental information, the implementation has been compromised by a charging policy which inhibits access, and is now the subject of a formal complaint being considered by the Commission. Moreover, in a number of key areas, such as industrial effluents discharged to sewers, drinking water quality and contaminated land, the public still do not have facilitated access to the environmental information they need.

The inadequacy of Freedom of Information regulation in the UK, particularly with respect to the activities of the private sector, reflects the general shortcomings of an approach based on a voluntaristic and anti-regulatory ideology. It creates a critical constraint on sustainable development.

The chemical release inventory

Both the value of information provision as an environmental protection measure and the weakness of UK policy in this respect are demonstrated by the case of the Chemical Release Inventory (CRI) (DOE 1992b) which would improve the availability of data relating to polluting emissions from industry.

In the US the Toxic Release Inventory (TRI) obliged companies to submit an annual list of the quantities of 322 hazardous chemicals they had 'released' into the environment. This induced a massive shift in regulatory power away from government and industry and towards the public, with the result that toxic emissions were substantially reduced. US Environmental Protection Agency administrator William Reilly declared that: 'The impact of TRI has far exceeded our expectations as a tool for improving environmental management [and that TRI data] should be considered to be among the most important weapons in efforts to combat pollution' (USEPA 1988, p. xxi).

Even in the absence of appropriate regulation, faced with citizens' demands, some companies will offer information on a voluntary basis. However, it is the company, not the community, which decides what particular information is released, and the information released is not necessarily in a form comparable with that released by other companies. In fact, only a few companies voluntarily disclose information. Surveys of toxic releases from chemical and other heavy industries reveal pervasive industry unwillingness to provide data unless forced to do so by law.

Out of 43 companies that operate in both the UK and the US and which were currently reporting release data in the US, six excused themselves on the grounds that their UK facilities were such that they would not have to report under the US laws. A large proportion (20) simply refused to provide the data. Where firms did volunteer information on releases, it revealed that a few large European facilities are releasing larger quantities of some toxics into water than the whole of US industry put together (Friends of the Earth 1992c).

This widespread withholding of data on chemical emissions contrasts markedly with the chemical companies' claims to be relatively open with the public and their supposed commitment to voluntary reductions in emissions of toxic substances. Although a Chemical Release Inventory has now been introduced, it does not provide company-specific information on emissions and only covers releases of the relatively small number of substances which require special authorisation from HMIP (FOE 1994b).

Implementing sustainability planning

Planning mechanisms

Target setting   Probably the single most fundamental test of any approach to achieving sustainable development is whether it includes targets for the maintenance of environmental capacity, based on a scientific understanding of the limits and tolerances of the biosphere and, in uncertainty, upon the precautionary principle. Targets are only the beginning of the process, and must be carefully defined and quantified. As the executive director of

UNEP has argued: 'What are needed now are specific commitments to take specific actions, over specific periods of time, with the costs calculated, the sources of funding identified, and a clear indication of who will be doing what. Nothing less will suffice.' (Tolba *et al.* 1992, p. 822).

In the Strategy, the Government claims that it has accepted this viewpoint: 'The UK believes that effective national strategies of this kind, containing real commitments and targets and substantive measures to achieve them, are essential in order to make progress on the problems affecting the environment of the whole world' (HMG 1994a, para. 7). But in the full text the message is less clear: 'Increasingly, it *may* be necessary to develop more specific environmental objectives or targets, for different environmental media or sectors of the economy' (HMG 1994a, para. 3.18, emphasis added).

And even this seems a rhetorical statement. In content, the Strategy notably lacks such targets, and the government is still prevaricating on the establishment of quantified and timetabled targets for environmental policy (FOE 1994a). In biodiversity protection, for example, the only targets put forward by the biodiversity action plan were suggested by NGOs and included for illustration only!

Targets can be set at different scales and levels. Most broadly, these may be limits to the consumption or erosion of environmental resources relating to the state of the environment, such as targets for an appropriate concentration of stratospheric ozone. But it is more common for targets to be set which relate directly to the pressure being exerted on that environmental resource (e.g. the emissions of ozone depletants) or indeed related to the policy response to the problem (e.g. reduced production of ozone-depleting substances). At this level, the ability to manipulate the outcome through policy tools is increased, but the degree of flexibility can be curtailed, reducing the likelihood of innovation and leading to less efficient solutions. The appropriate level of target varies between activities and sectors.

But without targets the allocation of environmental resources cannot be achieved. Such planning is necessary given the inability of the market to allocate certain resources effectively. Environmental resources are particularly hard to allocate as they are exceptionally difficult to value monetarily, and even where such values can be applied, working markets are difficult to establish. This does not prevent the use of market mechanisms to help meet targets, but does mean that such mechanisms cannot be relied upon to determine the appropriate level of environmental degradation.

Similarly, some mechanisms that are appropriate for target setting – such as the calculation of critical loads – are less appropriate as a tool for implementing change. For example, Germany has successfully exceeded its gap closure emissions reduction targets through the application of the principle of Best Available Technology (rather than seeking to reflect critical loads directly in pollution control regulation), with commensurate environmental benefits (Scharer 1993). Although the financial costs significantly exceed the costs incurred in the UK, the programme has provided employment for skilled workers from the shipbuilding industry who were being made redundant, thus saving costs that would otherwise have been incurred in welfare payments and providing substantial societal benefits.

Targets are already set in UK public policy at all scales; for example, as housing allocations, the inflation rate and morbidity rates for certain diseases. It is generally accepted that in such cases the market will not necessarily meet the public interest and need.

Under sustainability planning such quantified, timetabled targets would be more widely used, and defined by different criteria. For example, targets for pollution control would go beyond emission standards. Although these reduce unit emissions, in those cases where there is an increasing number of sources, such an approach can fail to absolutely control the degradation of the environmental resources. Nor can such an approach alone take the sensitivity of the receiving environment to that pollutant into account. Ambient standards, tailored to the sensitivity or vulnerability of the specific area, are the minimum necessary. Exposure levels of humans or the environment must take account of variation in the environment. For example, targets set to protect human health should reflect the response of the most vulnerable groups – for example, the one in seven with depressed cholinesterase levels with respect to organophosphate pesticides (Whittaker 1986). However, the lack of scientific certainty in the determination of critical target levels requires a strong precautionary approach. Thus, for example, pollution prevention should continue to take precedence over 'dilute and disperse' mechanisms. Equally, zero emissions remains the only safe target for persistent, toxic or bio-accumulative substances.

In other areas the government has operated a system of implied targets based on forecast demand. Such targets accept continuing environmental degradation without regard for sustainability constraints. For example, the road-building plans of the Department of Transport, based on forecasts of traffic growth, accept continued increases in carbon dioxide emissions, despite the recognised need to reduce such emissions. The government appears to be moving away from this approach in planning for road infrastructure and minerals supply, stating that such forecasts no longer constitute targets for policy (HMG 1994a). However, in neither case has the government yet reformed the policy implementation mechanisms which, in practice, translate the forecasts into *de facto* targets.

The process of 'sustainability planning' is fundamentally different from such 'trend-planning'. Indeed, in many cases the trend needs to be broken if we are to return to being within environmental capacity instead of, in effect, living beyond our environmental means. A target-setting process is being developed at a national level in the Netherlands where the National Plan is based on scenario projections, and sectoral targets are to be set to acceptable levels of environmental impact. In the UK, political resistance to target setting reflects factors such as market ideology and a lack of political will for public expenditure – obscured by often misplaced concerns over the financial costs or the likely burden on international industrial competitiveness, and consequential increased unemployment. The UK's failings are clearly demonstrated by policy in the critical field of climate change. Global warming has been studied in some detail and the sustainability constraint involved has been translated into rough global targets – most importantly for at least a 60% cut in emissions of carbon dioxide (SWCC 1990).

In the UK, the government's Programme for Climate Change fails to make any commitment to control carbon dioxide emissions beyond the year 2000. This is despite acknowledgement that: 'it seems likely that countries will need to take on new commitments [beyond 2000] if the Convention objective is to be achieved.' (HMG 1994b, para. 10.29).

Nor are adequate resources committed, although delay in necessary action will add to the costs of meeting future goals for carbon dioxide emission reduction. The only measure in the whole programme which will have a guaranteed sustained effect is the very modest planned improvement to the energy efficiency standards required in new buildings. The other measures proposed, particularly in the transport sector, largely deny this longer-term perspective, especially with respect to the implications of traffic growth. In this regard, the government is failing to implement the precautionary principle.

In the shorter term, the emissions programme is more robust, with two-thirds of the programme savings based on new taxes, extended commitments for existing measures to secure renewable energy development and public sector energy efficiency improvements, and new or improved energy efficiency standards for electrical appliances and buildings. However, the remaining third is less certain.

The largest doubt in the Programme is the Energy Saving Trust, a body established to deliver demand side management programmes, funded principally by the electricity and gas utilities. This is expected to deliver 2.5 million tonnes of carbon (MtC) (or one-quarter of the total target) by the year 2000. The Trust has estimated that this will require it to spend £1.5 billion between now and 2000. However, only just over £100 million of this funding has been secured, principally from electricity companies. The government has yet to determine how the remaining £1.4 billion will be secured. Unless the electricity and gas regulators establish an effective funding package, the Energy Saving Trust will fail.

The transport sector, responsible for 70% of the projected increase in carbon dioxide emissions, is expected to deliver only 25% of the savings required to meet overall targets. The most concrete measure in this sector is a commitment to increased road fuel duty by 5% in real terms each year until 2000. The likely impact of this is to slow the predicted increase in emissions from road traffic from 7 MtC to 4.5 MtC. However, the principal effect of the road fuel duty increase can be expected to be to encourage greater use of more fuel-efficient vehicles, rather than cut travel distances or curb traffic levels.

Computer modelling for Friends of the Earth by Earth Resources Research dramatically underlines the trade-off faced by policy-makers between technological improvements and levels of car use (Holman 1991). The level of car traffic growth that would be consistent with an interim target of a 30% cut in carbon dioxide emissions from car traffic by 2005 was calculated. Assuming rapid introduction of new technology, the target reductions in carbon dioxide were incompatible with both high and low NRTF forecasts. If fuel efficiency were improved at a slower rate, in line with improvements in recent years, car travel would have to be reduced if the target was to be met. In the Strategy, the government has admitted that: 'the impact of ever rising levels of transport on the environment is one of the most significant challenges for sustainable development. At the heart of this challenge lie the continuing forecasts for long term growth in demand for transport, together with the wide variety and level of associated environmental impacts' (HMG 1994a, para. 26.1).

It has also conceded that: 'if people continue to exercise their choices as they are at present and there are no other significant changes, the resulting traffic growth would have unacceptable consequences for both the environment and the economy of certain parts of the country and could be very difficult to reconcile with overall sustainable development goals' (HMG 1994a, para. 26.17).

Road transport currently contributes 18% of UK carbon dioxide emissions, but Department of Transport pollution forecasts predict a 43% increase in carbon dioxide emissions by 2010 (over a 1990 base) (Spackman 1992).

The policy implications are clear, as stated in the Strategy: 'Government as a whole should act to reduce environmental impacts and influence the rate of traffic growth' (HMG 1994a, para. 26.44). However, the Strategy does not introduce targets and timetables to control accelerating car use. Despite the increases in road fuel duty and new planning policy guidance on transport, the failure to set a limit for traffic growth fatally undermines the government's strategy for the transport sector.

Continuing growth in traffic will compromise the UK's ability to meet even its limited objectives for CO_2 emissions. Moreover it will continue to impact on air quality and on habitats, both directly and indirectly through road-building. Road-building to cater for continued traffic growth is especially controversial and environmentally damaging. Yet the government still questions *whether* road-building should be curtailed: '*If* it were no longer acceptable to build some roads, prices and physical management measures would be the best way to ration the limited resource' (HMG 1994a, para. 26.33, emphasis added).

In the case of measures to combat climate change, financial cost is not a constraint to action. Financial benefits to companies and the economy are being foregone as a result of limited action. In its consultation process on the Programme, the government itself identified a further 5 MtC of cuts which would be cost-effective, on top of the 10 MtC target. Friends of the Earth's input to the government's consultation on the Climate Programme identified a further 20 MtC of highly cost-effective savings by the year 2000 without including the transport sector (Karas *et al.* 1993). Further savings are possible from transport, depending not on cost-effectiveness, but on government willingness to switch funds from road building to public transport provision. A UK case study for the European Commission by the government's own Energy Technology Support Unit identified cost-effective potential to make a 30% cut in carbon emissions by 2005.

Most countries are still far from adopting targets in domestic policy and from implementing policy measures to meet this target. The fear of a burden on industrial costs is discouraging unilateral action, but in several countries has stimulated a range of revenue neutral proposals for carbon taxes, with compensatory reductions in labour taxes, described as 'ecological tax reform' (Weizsacker and Jesinghaus 1992). Modelling of such policies at the EU level indicates that such an approach would add to employment (Majocchi 1994).

This analysis leads inexorably to the conclusion that the slowness of government action does not depend on its stated concerns, but on other political and financial imperatives.

Demand management   To meet targets, supply side and efficiency measures may not prove adequate – as in the case of transport greenhouse gas emissions. In this case policies and measures to manage demand are required, such as land-use planning measures which reduce the need to travel (McLaren 1993a). In such ways 'sufficiency' strategies can be introduced and damaging trends fuelled by growth in population or GNP can be reversed.

However, here there is a clear risk that without careful introduction and positioning of such measures, they will prove politically unpopular with the public.

EIA and SEA   Our ability to control resource degradation and depletion rests upon effective measurement and prediction of the actual impacts of proposed policies and projects on environmental capacity. Environmental Impact Assessment at strategic and project levels must then lead to changes in project implementation. EIA currently fails to make adequate predictions even of the impacts of a *project*. It has yet to become an iterative process in which the actual impacts of those projects permitted to proceed are monitored and the results used to alter the conditions of operation where appropriate, and also made available for the preparation of other environmental impact statements (McLaren 1993b; Wood and Jones 1991).

Nor does EIA always consider alternative sites, processes, engineering options, demand-management solutions and the do-nothing option. Developers are not required to assess different ways of meeting the needs of the general public for their 'product', through substitution or even demand management.

At present, the information provided in the EIA is unlikely to be adequate to determine whether the development is sustainable because it fails to distinguish effectively between impacts on different groups, including future generations; and because the concentration of impact in time or space can be responsible for the loss of particular elements of the minimum environmental capital stock.

For example, despite EIA being required for afforestation in sites of special scientific interest (SSSIs) (Forestry Commission 1988), such sites are still being damaged by forestry activities. Analysis of NCC loss and damage statistics reveals that between 1986 and 1990, 38 sites were damaged in a total of 41 incidents, including three of inappropriate planting and six of drainage. The inadequacy of forestry EIA is partly because the competent authority for such assessment is the Forestry Commission, which, through the Forestry Enterprise, is responsible for much new planting.

The environmental impacts of the road schemes are appraised differently. The National Roads Programme is still designed principally to facilitate forecast traffic growth, and in several ways encourages such growth. The forecasts inform the selection of schemes which are then evaluated using a cost–benefit model in which most of the benefits are composed of expected time savings. Environmental impacts are assessed separately under procedures set out in the Manual of Environmental Assessment (MEA). Although a new MEA has recently been published (Department of Transport *et al.* 1993), past schemes and many of those still in the pipeline have been approved using a wholly inadequate system which, for example, excluded carbon dioxide. Nor does the new MEA guarantee improved implementation of EIA. Criticisms such as those levied by Treweek (1993), i.e. that ecological survey has been inadequate, could still be valid.

Most importantly, there is no system of strategic EIA of policies and programmes. Proposals for such a system are stalled at the EU, partly as a result of UK opposition, while policy appraisal initiatives in the UK have proved partial and ineffective.

Scientific certainty and the precautionary principle Sustainable development depends on the precautionary principle. All too often we do not know whether certain actions will seriously or irreversibly damage the environment on which we all depend, or what the appropriate target for environmental capacity should be. Our only effective defence in the face of such uncertainty is to apply the precautionary principle, and give the benefit of the doubt to the environment. This principle is enshrined in European Law in Article 130r(2) of the Single European Act (as amended by the Maastricht Treaty). The precautionary principle is clearly set out in the Bergen Declaration: 'In order to achieve sustainable development, policies must be based on the precautionary principle. Environmental measures must anticipate, attack and prevent the causes of environmental degradation. Where there are threats of serious or irreversible damage, lack of full scientific certainty should not be used as a reason for postponing measures to prevent environmental degradation' (1990 Bergen Ministerial Declaration on Sustainable Development in the ECE Region).

There is substantial industry and political opposition to the precautionary principle, as a result of at least two ideological factors. First, there is the tradition of a rational scientific basis to policy. Secondly, interference by the state in individual rights and freedoms has been strongly resisted, such that to remove such rights, it is expected that there must be *proof* that they damage other interests of recognised importance. The precautionary principle challenges this on the grounds that to delay, awaiting proof, may result in irreversible or much greater damage to environmental interests.

International agreements, Agenda 21 and national sustainability plans

As noted above, redistribution is the critical issue embedded in sustainable development at an international scale. This was central to debate at the Earth Summit: in particular, Southern countries were determined to obtain commitments from wealthy Northern nations to help meet the predicted US$600 billion per annum costs of implementing Agenda 21 (UNCED 1992).

But Agenda 21 is entirely voluntary and exhortative. Neither the agreement nor the process set up by the Commission for Sustainable Development (CSD) has any power to ensure international redistribution. At most, the CSD can seek to embarrass governments into further commitments or progress. These weaknesses are reflected in the UK's performance in the critical area of development aid.

The UK government's claimed achievements (HMG, 1994a) are dwarfed by the continuing abuse of aid finance to benefit UK business and political interests. The government is particularly proud of the UK's achievements on focused aid to support Agenda 21: 'In 1992–93, ODA had 179 projects under way whose principal objective was one of the Earth Summit priorities [forestry conservation, biodiversity, energy efficiency, population planning and sustainable agriculture]; these spent nearly £42 million, an increase of 21% over 1991–92. A further 162 projects included these priorities as a significant objective' (HMG 1994a, para. 28.34).

However, at £2100 million, the total UK aid programme budget for 1992/3 (ODA 1993) constitutes only 0.31% of GNP, falling far short of the longstanding UN target for aid of 0.7%, repeated by Agenda 21. The government has accepted this target, but has yet to set a date for meeting it.

In 1992 £28 million of UK aid was given to forestry projects in developing countries. But this figure was exceeded fivefold by the revenue received by the exchequer in VAT on imported tropical timber, none of which was produced sustainably (FOE 1992a, 1993).

Not only is there too little aid, but the quality of much UK aid is also questionable. Even if the £100 million provided by the UK for the Global Environment Facility is added to the £42 million on Earth Summit priorities, this forms only a fraction of the total budget. More money is spent on inappropriate projects. The poor quality of much UK aid is highlighted by the highly controversial Pergau hydro-electric scheme, personally sanctioned by the Prime Minister and Foreign Secretary against the explicit advice of the ODA's Permanent Secretary. This single project commits more UK tax-payers' money (albeit over 14 years) than all the 179 projects cited in the Strategy. The £234 million Pergau project was deemed a 'very bad buy' when the Overseas Development Administration assessed it (NAO 1993).

The environmental impacts of the project may also be significant and negative. The best available information indicates that the project may threaten one of the world's rarest and most endangered species, the Sumatran Rhinoceros, as well as other rainforest wildlife (ERL 1991). The project went ahead not because it was a development priority, but because the government wished to protect British commercial interests (Foreign Affairs Committee 1994).

Pergau may not be typical in every respect, but is representative of a series of large infrastructure projects funded under the Aid and Trade Provision, in that environmental assessment was rushed and limited in scope, the development benefits are questionable, and commercial interests were paramount.

Similar interests, but at a larger scale, inform multilateral development lending. World Bank lending is also predominantly directed at large infrastructure projects that create work for Northern companies (Chatterjee 1994; Rich 1994). Increasing proportions of lending are being targeted directly at the private sector through the International Finance Corporation. An increasing proportion of lending to governments is conditional on the adoption of structural adjustment programmes which restructure southern economies in ways which create markets for Northern manufactured exports and generate commodity and raw material exports. This process is reinforced by the GATT agreement where the interests of Northern companies have taken precedence over consideration of the potential environmental and developmental implications. In effect, the environmental damage of raw material and commodity supply is being exported from the North to the South.

However, Southern countries acting in concert have begun to exert some power. In intergovernmental debate over the Global Environment Fund (GEF), agreement has been reached to give recipient countries as much influence over the operation of the GEF as donors. However, the GEF represents only a tiny fraction of international capital flows, and the net flow of capital from debt financing and repayment still favours the North. The weaknesses of international policy are a critical constraint on sustainable development. Without strong and binding international agreements the export of damaging development cannot be halted, and equitable access to environmental resources cannot be achieved.

The constraints on sustainability planning

This concluding section reviews the different constraints which prevent effective sustainability planning in the UK.

Institutional

The institutional boundaries of systems for planning different land-based resources are limiting the effectiveness of the government's attempts at planning for sustainable development, despite initial attempts to achieve sustainable development through the land use planning system (DOE 1990).

In transport too, the separation of responsibilities is problematic. Local authorities are forced to include all Department of Transport highway schemes in their plans even where they run counter to locally agreed transport and planning policies. A survey by Friends of the Earth showed that many authorities were attempting to develop traffic restraint policies which could be seriously compromised by the provision of new trunk roads (Higman 1991). This problem is exacerbated by the failure of the Department of Transport to fund major public transport schemes in cities as part of the package funding initiative in which local authorities can bid for money for public transport as part of a 'package' including money for roads (Cook and Davis 1993).

The nature of regulatory systems has a strong bearing on the environmental standards achieved, while the powers given to regulators can vary significantly (OECD 1987; Pearce and Brisson 1993; ENDS Report 1993, 1994b).

However, all of these contributing institutional factors are essentially reflections of political constraints which are considered below.

Scientific/technical

The constraints imposed on our ability to achieve sustainable development by the limits to our knowledge cannot be ignored. Scientific work is needed on a range of issues such as pollutant interactions, epidemiology of pollutant impacts, indicators of critical loads and adaptive processes. Indeed basic data are often lacking regarding the size and nature of environmental resources. Technological developments will also contribute, ranging from habitat restoration techniques to product design.

But these are rarely limiting. Technological development is essentially a dependent variable, and appropriate policies can be expected to drive effective innovation. Scientific limitation means that policy approaches must rely more heavily on the precautionary principle.

But scientific and technological constraints are often represented as limiting factors by political and business interests.

Economic

The expectation that policies for sustainable development will be expensive is a common thread in government and business statements on the issues (HMG 1994a; Schmidheiny 1992). However, as noted earlier, this position cannot be substantiated. There are more fundamental economic constraints. In particular, the inability of the present economic system to distinguish between needs and wants compromises attempts to reduce our impacts on the environment while continuing to fulfil needs. The centrality of needs to our understanding of sustainable development suggests that this problem, when put in the context of government ideology about particular economic systems, merits much more detailed examination.

One aspect of this problem is revealed by attempts to place monetary values on the environment so that environmental costs can be incorporated into market mechanisms. It is debatable whether such valuation is technically possible, or morally acceptable; and some environmental economists would see our inability to derive accurate values as a serious constraint. However, if it is accepted that targets for environmental quality can be set politically, market mechanisms such as taxes can still be used to achieve these goals without attempting to value the environment in monetary terms.

Political and ideological

Examination of the political constraints provides more explanation of why progress is slow. The dominant political ideology of individualism and free markets is reflected in the evolution of the GATT at the international level and in many aspects of government policy at the national level such as privatisation and deregulation.

Voluntarism

Agenda 21 as a whole is an exhortative document but it recognises that a voluntary approach is not adequate to fulfil its aims at a national level: 'to effectively integrate environment and development in the policies and practices of each country, it is essential to develop and implement integrated, enforceable and effective laws and regulations' (UNCED 1992, para. 8.14).

But throughout the UK Strategy the voluntary approach forms a constant sub-text for action at individual and corporate levels, particularly through the involvement of sectoral business associations. However, its suitability as a policy tool is always assumed and never openly appraised.

It is increasingly clear that industry cannot deliver environmental targets through a voluntary approach. This is illustrated dramatically by the fact that the CBI's Environment Business Forum has not even been able to meet its membership target of 1000 firms. In fact, less than 200 bona fide businesses appear on the Forum's latest membership list. This has been attributed by the CBI to the unwillingness of companies to publish environmental reports (ENDS 1994a).

Despite the statement in the third year report on 'This Common Inheritance' that the Government 'is determined to maintain environmental standards' (HMG 1994b, para. 6) there is no guarantee that the general power to abolish regulation by ministerial powers sought under the deregulation bill will not be used to weaken existing environmental regulation.

Even industry supports effective regulation on environmental matters. The summary of the Department of the Environment's consultation meeting with industry reported a 'key point' that: 'legislation was necessary to secure minimum standards . . . a synergistic mix of instruments, with a regulatory base, was necessary' (DOE 1993b, para. 2c).

After the launch of the Strategy, John Cridland, Director of the Environment Unit at the Confederation of British Industry stressed that business has a voluntary role to play above and beyond regulation, not instead of regulation (Cridland 1994). But the government makes no commitment to a strong regulatory framework in the Strategy. The strength of this ideology is revealed by the approach taken to packaging waste. The packaging industry was repeatedly exhorted to introduce a voluntary scheme to reduce packaging waste volumes, and even when the industry decided that a legislative foundation would be required, ministers are equivocating (DOE 1994).

Regulation and costs

Overall, the premise that current UK environmental regulation imposes a burden on the business community may need to be challenged. The law is often weak, there is relatively little enforcement action and the regulators are themselves under-resourced. In fact, the environmental and social costs of failing to provide the appropriate regulatory framework are likely to far outweigh any financial burden involved. Indeed, there is evidence to suggest that, far from imposing a cost burden on business, strong environmental regulation can be financially good for business. Clearly strong environmental regulation can benefit firms manufacturing environmental technology, but firms subject to regulation can benefit too.

Environmental regulation can stimulate innovation for compliance which can provide both a competitive advantage and new markets (Ashford and Heaton 1979; OECD 1985; Willson and Greeno 1993). A study of innovation in five countries found that policies aimed specifically at stimulating innovation showed no association with successful projects, while regulations were positively related to innovation performance (Allen et al. 1978). Firms can also increase profitability through cleaner technology which cuts costs for raw materials, energy and waste disposal. Such process efficiency improvements can pay back the investment rapidly (Huisingh 1988).

Evidence also suggests that change in corporate practices is most effectively triggered by appropriate regulation. In a 1991 survey of large trans-national corporations by the United Nations, 57% of respondents cited home country legislation as the most influential factor provoking company-wide changes with respect to environmental policy and programmes (UNEP 1991). This is reflected in the UK. Of the forces contributing most to companies' concern about environmental pressures the need to comply with UK and European Community legislation was cited as the main reason by 36% of directors, social responsibility and corporate citizenship by

26% and market forces by just 23% (Institute of Directors 1993), indicating that self-regulation is not in line with business reality.

At an international level there is also consensus on the need for appropriate regulation. The United Nations Environment Programme (Industry and Environment Programme) states that:

> environmental laws and regulations are required to ensure that all companies equally conduct their activities by at least basic environmental standards. If laws and regulations are not applied fairly, systematically and effectively, environmental objectives will not be met and governments risk losing credibility. (Tolba *et al*. 1992).

'Towards Sustainability' states that:

> various legislatively-based rules, standards and procedures [should] be applied to the different stages of the authorisation–production–appraisal chain so as to create a self-perpetuating inducement to progressively apply ever-improving standards. (CEC 1992a, p. 29.)

While Agenda 21 states that:

> Laws and regulations suited to country-specific conditions are among the most important instruments for transforming environment and development policies into action, not only through 'command and control' methods, but also as a normative framework for economic planning and market instruments. Yet, although the volume of legal texts in this field is steadily increasing, much of the law-making in many countries seems to be ad hoc and piecemeal, or has not been endowed with the necessary institutional machinery and authority for enforcement and timely adjustment. (UNCED 1992, para. 8.13.)

The inadequacies of the voluntary approach are highlighted by the case of energy efficiency. The government has acknowledged that the UK wastes 20% of the fossil fuels and electricity it uses (DOE 1993a, para. 3.25), indicating that a substantial and cost-effective drop in consumption could be achieved by improving energy efficiency and energy saving. The government emphasises 'voluntary action' by individuals and businesses in the absence of any new government initiative, beyond, perhaps, information campaigns.

As Karas *et al*. (1993) argue, such an approach hides three largely unexamined assumptions: first, that improving energy efficiency appears to be in the interest of the individual householder or business; secondly, that action by the public can be brought forward by a higher level of government exhortation; and, thirdly, that the only alternative to 'voluntary action' is intervention and coercion. These beliefs are fundamental to the government's approach but do not stand up to examination.

Although there are 'cost-effective' measures which are in the interests of consumers to undertake, there are also barriers and obstacles in the 'market in energy services' to the efficient allocation of resources (Jackson 1992; Roberts 1992; Karas *et al*. 1993). These barriers combine to cause individuals and businesses to divert investment into less energy efficient options than would be deemed optimal from the perspective of national economic interests. In terms of household and business cash-flows it is often easier to pay fuel bills than to find and finance the capital investment required to improve energy efficiency. Moreover, with energy utilities controlled by regulations which encourage sales expansion, the overall balance of propaganda materials in the energy market is driving consumption rather than conservation.

Consumers are currently getting the wrong signals in the energy market. The correct 'signals' are unlikely to appear spontaneously without government action. But it is not a question of the government having to become heavily interventionist in the energy market – it already is, but its influence is directed contrary to energy efficiency. Government policy already sets energy standards for new buildings; establishes the regulatory framework which controls the behaviour of electricity and gas companies; decides whether VAT must be charged on energy efficient building improvements; grants planning consent for new power stations and pays for public information campaigns on energy efficiency.

Nor is it a question of the government necessarily needing to find new money to pay for the measures to improve energy efficiency. It is more a question of bringing forward policies which have the effect of redirecting the expected expenditure and investment so that it produces energy savings rather than energy supply.

Economic instruments

The government's market ideology is also reflected in its repeatedly stated, if unfulfilled, preference for economic instruments.

The Strategy states that: 'in environmental policy, the commitment is to make use of economic instruments where possible, rather than regulation' (HMG 1994a, para. 3.21).

Such a commitment prejudices any assessment of the effectiveness of a particular policy tool in achieving high environmental quality in specific circumstances. The government should make a commitment to using the instrument that is most effective in achieving gains in environmental quality.

The government's political preference for economic instruments is not new but has yet to be reflected substantially in policy. Nor have those instruments put in place been fully appraised. Unleaded petrol is often quoted as the flagship example of the success of economic instruments, but this is questionable. Unleaded petrol has been much cheaper than leaded petrol for several years, but has only recently exceeded 50% of UK petrol sales (Warren Spring Laboratory 1993). A substantial proportion of drivers easily able to take advantage of this price differential have therefore failed to do so. More generally, the effectiveness of economic instruments is compromised by market failures (such as barriers to investment or lack of information).

Privatisation

The political pressure behind the free-market economy has meant that privatisation of previously state-run operations has featured heavily in recent UK legislation. While in theory privatisation can benefit the environment by improving the efficiency of the newly privatised enterprises or releasing new money for investment, in fact the environmental record of privatisation in the UK has been poor. The water industry is still behind the targets set by European legislation, and indeed has argued that EU water quality standards should not be met. The energy utilities have embarked on marketing drives to increase sales, rather than focusing on energy efficiency (OFFER 1993). The view that privatised transport systems will be more efficient and that 'it is not the Government's job to tell people where and how to travel' (HMG 1994a, para. 26.17) have combined to threaten public transport availability and use. Deregulation of buses has contributed to a significant decline in bus use (Stokes *et al.* 1990). Rail privatisation may result in a similar cherry-picking exercise on profitable routes with declining use overall. A common thread runs through these examples. In all cases the government has put the maximisation of financial returns to investors (corporate and individual) and to the Treasury above creating a regulatory framework for the new industry that will protect environmental interests. But this is not fundamental to privatisation. For example, in many American states privatised energy companies operate under a duty to undertake 'least-cost planning' which favours investment in energy efficiency (Flood 1992), although it may not maximise profits.

This is one way in which government policy reflects the short-term interests of particular lobby groups – in this case the corporate investors. Similarly aid policy brings benefits to British companies, and in some cases the choice of projects is strongly influenced by such interests (Foreign Affairs Committee 1994). And many of the companies with such interests are Conservative party donors (Dilworth and McLaren 1994). In transport too, the influence of the 'roads lobby' can easily be discerned (Hamer 1987), and its strength has contributed to conflicts between government departments. Recent transport policy has exhibited two divergent strands. On the one hand, the Department of the Environment (1992a) advocates the use of the development planning and control systems to reduce the need for travel and to encourage less damaging modes. On the other, the Department of Transport (1992) seems to regard the integration of transport and land use as no more than the extraction of development gain to pay for road improvements. The preparation and publication of PPG13 has revealed continued conflict between these departments. The Department of Transport has sought to ensure that potential congestion on trunk roads is considered by planners, but resisted any suggestion that planners should be able to challenge Department of Transport road schemes which might create undesirable development pressures.

Incrementalism

The power of vested interests is also reflected in an incremental approach to policy development. Environmental measures have been 'bolted onto' an existing policy framework in a way which has rarely been proactive. Policies have reacted to environmental problems as they have arisen or become subjects of public concern. Rather than setting targets based on the quality of the environment (or environmental capacity), where targets have been set, more often they have been based on emissions levels. However, more recently, and particularly in the development of European Commission environmental policy, environmental quality standards have been introduced for some media. For example, in land use planning, areas have been identified where certain types of development will be discouraged. This may preserve some particular environmental assets, but cannot halt the ongoing deterioration of our environment through, for example, increasing ambient pollution levels from traffic.

In part this bolt-on approach is due to a belief amongst politicians that environmental protection will damage economic viability at the national scale. As Blowers (1992) suggests, states are facing the 'prisoner's dilemma'. Nationally as well as locally this discourages unilateral action on environmental protection. However, there is no evidence that economic disbenefits will arise, and indeed research indicates that investment in environmental protection will increase employment and economic growth (Meissner 1986; Jenkins and McLaren 1994).

The historical legacy of environmental planning is as a poor relation to economic policy. Policies and practices which further the conservation of environmental resources through demand management have yet to be widely developed. This is not unexpected, as such approaches run counter to conventional approaches to economic development. But supply management alone cannot be relied upon. In land use planning, where the system ensures that sites are available, for example for waste disposal, this depresses land values so that they do not reflect full environmental costs and therefore inhibits the development of less damaging alternatives. The demand-led system effectively gives a subsidy to polluting industry and damaging development through land provision. This then distorts attempts to impose full environmental costs through, for example, the pollution control system. Moreover, the demand-led approach has compromised policies to control land use to reduce transport needs (McLaren 1993b).

Conclusion

If sustainability planning is to be embraced then deep-rooted ideologies need to be redefined. The concept of the free market must
embrace a regulatory framework in the public interest. Economic
systems must be able to recognise and meet needs in every country. Individual rights must be constrained where to exercise them
would damage environmental interests; in other words, the right
to clean air must be as important as freedom of movement, for
example. These are far-reaching challenges, which merit further
investigation as well as demanding political action.

Postscript

Sustainability, Planning and the Future

SUSAN BUCKINGHAM-HATFIELD and BOB EVANS

> Market forces cannot produce sustainability because individual firms and consumers cannot act in the knowledge of what everyone else is doing. They therefore do not know what the combined result will be. Even those who wish to do so face a prisoner's dilemma. If an individual person uses their car less to cut down on CO_2 emissions, but nobody else does, global warming is unaffected and they lose out . . . . Only if society acts collectively to influence and constrain individual market decisions – by requiring firms to reduce pollution, and by making public transport a genuine alternative to private car use – can sustainability be achieved. (Jacobs 1995, p. 7.)

The policy goal of sustainability and the policy process of environmental planning are, in our view, inextricably intertwined. The argument that unfettered markets will somehow deliver a more sustainable society is unconvincing. On the contrary, as Jacobs points out, the central, core values of sustainability are collective in that, implicit in the definition of sustainability, is a commitment to a sharing of common futures and fates, and a preparedness to take decisions in the interests of as yet unborn generations. All this points to public action, to planning, to decision taking in the 'public interest', and it is for this reason that environmental planning must be seen as the way of achieving environmental sustainability.

However, in the call for a return to public, collective action, and for a reinstatement of the concept of 'planning' as a legitimate and necessary activity of government, it is necessary to retain a sense of

Environmental Planning and Sustainability. Edited by S. Buckingham-Hatfield and B. Evans.
© 1996 by John Wiley & Sons Ltd.

balance and a sense of history. It is commonplace to point to the environmental and ecological degradation which is associated in many people's mind with the former communist Eastern bloc countries. That these centralised command economies promoted massive environmental exploitation cannot be doubted, and although the reasons for this are complex and, in the context, perhaps understandable, it is not unusual for the conclusion to be reached that, in environmental terms at least, 'public planning' cannot work.

By the same token, the experience of post-war town planning and urban renewal in Britain has left a legacy of distrust. 'The planners' have been blamed, sometimes legitimately, for soulless urban developments, suburban wasteland estates and car-dominated town centres. By the 1970s, critics from the political left and right, developers, academics, journalists, community activists, farmers, landowners, developers and so on were uniformly critical of 'planning' and it was this widespread disillusion, as much as the Thatcherite project, which forced the gradual retreat of planning into a kind of semi-judicial process of private sector regulation.

So, the central point that we wish to make is this: environmental planning as a process to strive for sustainability is, in our view, an essential and necessary component of a society such as ours. It has to be recognised, however, that there is an urgent need to rebuild public confidence in the legitimacy of 'planning', and that this will not come easily. Too many 'ordinary people' have day to day experiences that have severely shaken any belief which they or their forbears might have had in the ability and sincerity of public planners.

And yet, the rebuilding of this confidence is an essential prerequisite for the sustainability project. As the chapters in this book have made very clear, questions of social equity, of ownership, and of democracy, are at the very heart of the notion of sustainability. Until there is a degree of public confidence in the validity of sustainability as a policy goal, and in the mechanism of environmental planning as a way of getting us there, little of substance will happen.

Of course, environmental education is essential to this process. There is a very real need to explain why certain environmental policies are necessary, and what the consequences of inaction might be. But there is also a need for those involved in the business of environmental planning – practitioners, academics, researchers, politicians, activists – to seek to rebuild public confidence through more open, tolerant, participatory policy-making, devoid of the spurious 'top-down' professionalism which characterised so much early town planning. The nature of the policy-making process must also change so that co-operation replaces competition between professionals and so that a genuine and sustained dialogue which informs policy-making might take place between 'experts' and informed citizens.

Most modern societies are characterised by inequality, subordination and domination, and these social and economic circumstances have inevitable environmental consequences. Those with economic and political power will tend to secure the best environments, and will usually be prepared to tolerate environmental exploitation and degradation as long as this is confined to localities which they do not need, value or frequent.

It would be naïve to pretend that a reconstructed and more democratic environmental planning will somehow overcome these realities. However, as we argued in Chapter 1, the concept of sustainability is a fundamentally political one, and although many environmental planners would prefer not to recognise it, 'the environment' is as politically contested as any other area of human activity and endeavour. Our hope is that this book will help stimulate this political debate as a contribution towards more effective and democratic action for a more sustainable environmental future.

References

Agarwal A and Narain S (1992) The fridge, the greenhouse and the carbon sink. *New Internationalist* 230, 10–11

Agyeman J (1988) Ethnic minorities – an environmental issue? *ECOS* 9 (3), 2–5

Agyeman J (1989) Black people, white landscape. *Town and Country Planning* 58 (12), 336–338

Agyeman J (1992) Who's ready for freedom of environmental information? *Town and Country Planning* 61 (11/12), 293–294.

Agyeman J and Evans B (Eds) (1994) *Local Environmental Policies and Strategies.* Longman, London

Agyeman J and Evans B (1995) Sustainability and democracy: community participation in Local Agenda 21. *Local Government Policy Making* 22 (2)

Agyeman J and Tuxworth B (1994) *Local Authority Environmental Publicity Guide.* Central Local Government Environment Forum/HMSO

Agyeman J, Warburton D and Wong J L (1991) *The Black Environmental Network Report.* BEN, London

Allen T J, Utterback J M, Sirbu M A, Ashford N A and Hollomon J H (1978). Government influence on the process of innovation in Europe and Japan. *Research Policy* 7(2)

Anand A (1983) Saving trees, saving lives: Third World women and the issue of survival. In L Caldicott and S Leland (Eds) *Reclaim the Earth.* Women's Press, London

Armstrong P and Armstrong H (1988) Taking women into account. In J Jenson, E Hagen and C Reddy, *Feminisation of the Labour Force.* Polity, Cambridge

Ashford N A and Heaton G R (1979) The effects of health and environment regulation on technological change in the chemical industry: theory and evidence. In C T Hill (Ed.) *Government Regulation and Chemical Innovation.* American Chemical Society Symposium Series No. 109. American Chemical Society, Washington DC

Association of Metropolitan Authorities (1990) *Changing Gear.* AMA, London

Ave G et al. (1994) *The Economic and Cultural Conditions of Decision Making for the Sustainable City.* Final Report to the Commission of European Communities prepared by the London School of Economics and the Politecnico di Torino.

AWRG (UK Acid Waters Review Group) (1989) *Acidity in UK Fresh Waters.* HMSO, London

Baas J M, Ewert A and Chavez D J (1993) Influence of ethnicity on recreation and natural environment use patterns: managing recreation sites for ethnic and racial diversity. *Environmental Management* 17 (4), 523–529

Bannister C (1990) Existing travel patterns: the potential for cycling. In *Cycling and the Healthy City*. Friends of the Earth, London, pp. 20–28.

Barr C J *et al.* (1993) *Countryside Survey 1990: Main Report*. DOE, London

Bayliss D H (1994a) State-of-the-environment reporting: a review. In I Fodor and G P Walker (Eds) *Environmental Policy and Practice*. Hungarian Academy of Sciences, Pecs, Hungary

Bayliss D H (1994b) State-of-the-environment reporting and planning for sustainability. Unpublished MA Thesis, University of Central England School of Planning, Birmingham

Bayliss D H and Walker G P (1992) Environmental index. *Energy Policy* 21, 3–4

Bayliss D H, Belicic T, Gourio L and Pappavassiliou V (1993) *Environmental Quality in Thessaloniki, Stoke-on-Trent, Erlangen and Rennes*. Department of Housing and Health, Stoke-on-Trent City Council, Civic Centre, Stoke-on-Trent

Bishop R C (1993) Economic efficiency, sustainability and biodiversity. *Ambio* 22 (2/3), 69–73

Blowers A (1992) Sustainable urban development: the political prospects. In M Breheny (Ed.) *Sustainable Development and Urban Form*. Pion, London

Blowers A (Ed.) (1993) *Planning for a Sustainable Environment*. Earthscan, London

Boulter R (1994) Reexamining the wheel. *Planning Week* 14 April 1994

Braidotti R, Charkeiwicz E, Hausler S and Wieringa S (1994) *Women, the Environment and Sustainable Development*. Zed, London

Braybrooke D (1974) *Traffic Congestion Goes Through the Issues Machine*. Routledge, London

Breakwell G M (1992) *Social Psychology of Identity and the Self Concept*. Surrey University Press, London

Breakwell G M and Canter D (1993) *Empirical Approaches to Social Representations*. Oxford University Press, Oxford

Breheny M (1991) The renaissance of strategic planning? *Environment and Planning B: Planning and Design* 18 (2), 233–249

Breheny M (1992) *Sustainable Development and Urban Form*. Pion, London

Buckingham-Hatfield S (1994) Popular concerns and the environmental agenda: on involving women in formulating local responses to Agenda 21. In I Fodor and G Walker (Eds) *Environmental Policy and Practice*. Hungarian Academy of Sciences, Pecs, Hungary

Buitenkamp M, Venner H and Wams T (Eds) (1993) *Action Plan Sustainable Netherlands*. Friends of the Earth Netherlands, Amsterdam

Bullard R D (1993a) Anatomy of environmental racism. In R Hofrichter (Ed.) *Toxic Struggles: The Theory and Practice of Environmental Justice*. New Society Publishers, Philadelphia

Bullard R (Ed.) (1993b) *Confronting Environmental Racism, Voices from the Grass Roots*. South End Press, Boston, MA

Camhis M and Fox S (1992) The EC as a catalyst for European Urban Networks. *Ekistics* 352/353, 4–6

Campaign for the Protection of Rural England (1993) *Sense and Sustainability: Land Use Planning and Environmentally Sustainable Development*. CPRE, London

Carson R (1962) *Silent Spring*. Houghton Mifflin, Boston, MA

Chatterjee P (1994) *50 Years is Enough! 50 Years of World Bank Development Projects*. Friends of the Earth, London

Chalmers A F (1978) *What is This Thing called Science?* Open University Press, Milton Keynes

Clayton A (1992) Cities and the environment – is urban living sustainable? *The Planner* 78, 21 Report of Summer School Proceedings, pp. 9–11

Collard A (with J Contrucci) (1988) *Rape of the Wild*. The Women's Press, London

Commission for Racial Equality/Royal Town Planning Institute (1983) *Planning for a Multi-racial Britain*. CRE, London

Commission for Racial Justice (1987) *Toxic Wastes and Race in the USA*. United Church of Christ, New York

Commission of European Communities (1990) *Green Paper on the Urban Environment*. COM(90)218, CEC, Brussels

Commission of European Communities (1991) *Europe 2000: Outlook for the Development of the Community's Territory*. CEC, Brussels

Commission of European Communities (1992a) Towards Sustainability. A European Community programme of policy action in relation to the environment and sustainable development. *Official Journal of the European Communities*, No. C138

Commission of European Communities (1992b) *Green Paper on the Impact of Transport on the Environment: A Community Strategy for 'Sustainable Mobility'*. Com92(46), CEC, Brussels

Commission of European Communities (1992c) *Treaty on European Union*. Council and Commission of the European Communities, Brussels

Commission of European Communities (1994a) *Background Report: The European Environment Agency*. ISEC/B6/94, CEC, Brussels

Commission of European Communities (1994b) *Community Initiative concerning Urban Areas (URBAN)*. COM(94)61 final, 2 March 1994, CEC, Brussels

Cook A J and Davis A L (1993) *Package Approach Funding: A Survey of English Highway Authorities*. Friends of the Earth and the University of Westminster, London

Coontz S and Henderson P (1986) *Women's Work, Men's Property*. Verso, London

Costello A, Vallely B and Young J (1989) *The Sanitary Protection Scandal*. Womens Environmental Network, London

Cotgrove S and Duff A (1980) Environmentalism, middle class radicalism and politics. *Sociological Review* 28, 333–351

Council for the Protection of Rural England (1993) *Sense and Sustainability*. CPRE, London

Council of Europe (1984) *European Regional/Spatial Planning Charter (The Torremolinos Charter)*. Council of Europe, Strasbourg

Council of Europe (1988/1992) *European Regional Planning Strategy*. Council of Europe, Strasbourg

Countryside Council for Wales (1993) *Annual Report*. Countryside Council for Wales, Bangor

Cridland J (1994) *Speech to ERM Environmental Forum, Sustainability: The First Steps*. Royal Society of Arts, 25 January

Cross M and Keith M (Eds) (1993) *Racism, the City and the State*. Routledge, London

Cutter S L (1995) Race, class and environmental justice. *Progress in Human Geography* 19(1), 111–122

Cyclists' Public Affairs Group (1994) *Trust Pedal Power: A Review of Transport Policies and Programmes for 1994/1995 with Regard to Cycling*. CPAG, Godalming

Cyclist's Touring Club (1992) *Cycling and Statistics*. CTC Occasional Paper No. 1, CTC, Godalming

Cyclist's Touring Club (1993a) *Cycle Policies in Britain: The 1993 CTC Survey*. CTC, Godalming

Cyclist's Touring Club (1993b) *The Greater Nottingham Cycle Route Project: A Case Study Report*. CTC Occasional Paper No. 2, CTC, Godalming

Dalla Costa M (1988) Domestic labour and the feminist movement in Italy since the 1970s. *International Sociology* (1), 23–34

Dankelman I and Davidson J (1988) *Women and Environment in the Third World.* Earthscan, London

Davies H and Gosling J (1994) *The Impact of the European Community on Land Use Planning in the United Kingdom.* Royal Town Planning Institute, London

Davies H et al. (1989) *Planning Control in Western Europe.* HMSO, London

Dawson J (1992) European city networks: experiments in transnational urban collaboration. *The Planner* 78(1), 7–9

De Beauvoir S (1968) *The Second Sex.* Random House, New York

De la Court T (1990) *Beyond Brundtland.* Zed, London

Department of the Environment (1990) *This Common Inheritance.* HMSO, London

Department of the Environment (1991) *Policy Appraisal and the Environment.* HMSO, London

Department of the Environment (1992a) *This Common Inheritance: First Year Report.* HMSO, London

Department of the Environment (1992b) *The UK Environment.* HMSO, London

Department of the Environment (1992c) *Urban Air Quality in the UK, First Report of the Quality of Urban Air Review Group.* Department of Environment, London

Department of the Environment (1992d) *Consultation Paper: Proposed Chemical Release Inventory.* DOE, London

Department of the Environment (1992e) *PPG12: Development Plans and Regional Planning Guidance.* HMSO, London

Department of the Environment (1993a) *The UK Strategy for Sustainable Development: Consultation Paper.* HMSO, London

Department of the Environment (1993b) Note of the Round Table Meeting to Discuss the UK Sustainable Development Strategy 12 August 1993.

Department of the Environment (1993c) *Making Markets Work for the Environment.* HMSO, London

Department of the Environment (1993d) *The Environmental Appraisal of Development Plans: A Good Practice Guide.* HMSO, London

Department of the Environment (1993e) *This Common Inheritance: Second Year Report.* HMSO, London

Department of the Environment (1993f) *PPG6: Town Centres and Rural Developments,* HMSO

Department of the Environment (1994a) News Release 553. John Gummer and Michael Heseltine congratulate industry on efforts to boost recycling of packaging. 28th September 1994

Department of the Environment (1994b) *This Common Inheritance: Third Year Report.* HMSO, London

Department of the Environment/Department of Transport (1994) *Planning Policy Guidance 13: Transport.* HMSO, London

Department of Transport (1989) *National Road Traffic Forecasts 1989 (Great Britain).* HMSO, London

Department of Transport (1991) *Evidence to House of Commons Transport Committee: Cycling.* HMSO, London

Department of Transport (1992) Press Notice 220. MacGregor unveils proposals to boost developer contributions to road improvements. 14 August 1992

Department of Transport (1994) *June Cycling Statement.* DTp, London

Department of Transport, Scottish Office Industry Department, Welsh Office and Department of the Environment Northern Ireland (1993) *Design Manual for Roads and Bridges,* Vol. 11. Environmental Assessment, HMSO, London

Diamond I and Feman Orenstein G (1990) *Reweaving the World.* Sierra Club Books, San Francisco

Dilworth A and McLaren D (1994) *Pergau and Other Stories.* Friends of the Earth, London

Dobson A (1990) *Green Political Thought.* Harper Collins, London

Douthwaite R (1992) *The Growth Illusion.* Green Books, Bideford

Duncan J and Ley D (Eds) (1993) *Place/Culture/Representation.* Routledge, London

East Hampshire Council (1993) *Towards Sustainable Development: An Environmental Strategy for East Hampshire.* April 1993

Edwards M (1992) Sustainability and people of colour. *EPA Journal*, Washington DC September/October, 50–51

Edwards R, Gee A and Stoner J (Eds) (1989) *Acid Waters in Wales.* Kluwer Academic, London

Ekins P (1992) *A New World Order: Grassroots Movements for Global Change.* Routledge, London

Elkin T and McLaren D with Hillman M (1991) *Reviving the City.* Friends of the Earth, London

English Nature (1993) *Annual Report.* English Nature, Peterborough

English Nature (1994) *Strategic Planning and Sustainable Development.* English Nature, Peterborough

ENDS Report 223 (1993) *Enforcement Suffers as HMIP helps Industry with IPC.* August, p. 9

ENDS Report 230 (1994a) *CBI bids to revive environment forum.* March

ENDS Report 234 (1994b) *HMIP steps up enforcement action.* July, p. 10

Environ (1994) *The Asian Community and the Environment – Towards a Communication Strategy.* Environ Research Report, Leicester

Epstein B (1993) Ecofeminism and grass-roots environmentalism in the United States. In R Hofrichter (Ed.) *Toxic Struggles: The Theory and Practice of Environmental Justice.* New Society Publishers, Philadelphia

ERIC (1995) *Local Agenda 21 Survey 1994/5.* Environmental Resource and Information Centre, University of Westminster/Local Government Management Board

ERL (1991) *Pergau Hydro Electric Project: Environmental and Socio-Economic Issues.* Report of a Consultancy Visit, Report to ODA

EU Presidency (1993) *Conclusions of the Presidency on the Informal Council of Ministers Responsible for Regional Policy and Regional Planning.* Liege, November 1993, Brussels

Evans B (1994) Planning, sustainability and the chimera of community. *Town and Country Planning* 63(4)

Evans B (1995) *Experts and Environmental Planning.* Avebury, Aldershot

Federal Ministry for Regional Planning, Building and Urban Development (1993) *Spatial Planning Policies in a European Context.* Bonn

Fischer F and Forester J (Eds) (1993) *The Argumentative Turn in Policy Analysis and Planning.* UCL Press, London

Flood M (1992) *Energy Without End.* Friends of the Earth, London

Foreign Affairs Committee (1994) *Third Report. Public Expenditure: The Pergau Hydro-Electric Project, Malaysia, The Aid and Trade Provision and Related Matters.* HMSO, London

Forestry Commission (1988) *Environmental Assessment of Afforestation Projects.* Forestry Commission, Edinburgh

Friend A M and Rapport D J (1991) Evolution of macro-information systems for sustainable development. *Ecological Economics* 3, 59–76

Friends of the Earth (1989) *Environmental Charter for Local Government*, FOE, London

Friends of the Earth (1992a) *Whose Hand on the Chainsaw? UK Government Policy and the Tropical Rainforests*. FOE, London

Friends of the Earth (1992b) Press Release, 28 March 1992. Report of opinion survey of 15 000 shoppers

Friends of the Earth (1992c) Press Release, 29 July 1992

Friends of the Earth (1993) *Forests Foregone*. Friends of the Earth, London

Friends of the Earth (1994a) Memorandum of Evidence to House of Lords Select Committee on Sustainable Development. Unpublished

Friends of the Earth (1994b) Press Release. Fundamental Flaws in Official Chemical Release Inventory. 7th September 1994

Friends of the Earth Europe (1994) *Sustainable Europe*. FOE Europe, Brussels

Fudge C (1992) *Urban Environment Group – from Green Paper to White Paper?* European Information Service, Issue 130, June 1992, pp. 5–7

Fudge C (1993) *Fragmentation and Progress – Urban Environment Expert Group Focuses on Sustainable Cities*. European Information Service, Issue 143, September 1993, pp. 11–13

Galbraith J K (1992) *The Culture of Contentment*. Sinclar-Stevenson, London

Gee D, McLaren D P and Crabtree T (1990) *Beyond Rhetoric: An Economic Framework for Environmental Policy Development in the 1990s*. Friends of the Earth, London

Gilman C P (1915) *Women and Economics*. G P Putnams Sons, London

Goldberg D T (1993) Polluting the body politic: racist discourse and urban location. In M Cross and M Keith (Eds) *Racism, The City and the State*. Routledge, London

Greed C (1994) *Women and Planning; Creating Gendered Realities*. Routledge, London

Griffin S (1978) *Women and Nature*. Harper & Row, New York

Haigh N (1989) *EEC Environmental Policy and Britain*. Longman, Harlow

Hajer M (1994) Ecological modernisation and social change. In S Lash, B Szerszynski and B Wynne (Eds) *Risk, Environment and Modernity: Towards a New Ecology*. Sage, London

Hamer M (1987) *Wheels within Wheels: A Study of the Road Lobby*. Routledge and Kegan Paul, London

Hams T (1994) Local environmental policies and strategies after Rio. In J Agyeman and B Evans (Eds) *Local Environmental Policies and Strategies*. Longman, London

Hansson C and Liden K (1984) *Moscow Women*. Allison & Busby, London

Harcourt W (1994) *Feminist Perspectives on Sustainable Development*. Zed, London

Harding S (1993) Rethinking standpoint epistemology: what is strong objectivity? In L Alcoff and E Potter (Eds) *Feminist Epistemologies*. Routledge, London

Hartsock N (1984) *Money, Sex and Power*. Northeastern University Press, Boston

Harvey D (1982) *The Limits to Capital*. Blackwell, Oxford

Harvey D (1989) From managerialism to entrepreneurialism: the transformation in urban governance in late capitalism. *Geographiske Annaler Series B* 71B(1), 3–18

Harvie C (1994) *The Rise of Regional Europe*. Routledge, London

Healey P (1991) Researching planning practice. *Town Planning Review* 62(4), 447–459

Healey P and Shaw T (1994) Changing meanings of the environment in the British planning system, *Transactions of the Institute of British Geographers* 19(4), 425–438

Healy R G (1987) State of the environment reports. *Journal of Planning Literature* 2, 262–272

Her Majesty's Government (1994) *Climate Change: The UK Programme*. HMSO, London

Higman R (1991) Assessing the new realism. *Surveyor*, 3 October

Hirsch J (1981) The apparatus of the State, the reproduction of capital and urban conflicts. In M Dear and A J Scott (Eds) *Urbanisation and Urban Planning in Capitalist Society.* Methuen, London

Hobsbawm E (1991) Dangerous exit from a stormy world. *New Statesman and Society* 4, 176

Hohl A and Tisdell C A (1993) How useful are environmental safety standards in economics? The example of safe minimum standards for the protection of species. *Biodiversity and Conservation* 2, 168–181

Hohn C F (1976) A human-ecological approach to the reality and perception of air pollution: the Los Angeles case. *Pacific Soc. Review* 19 (January)

Holman C (1991) *Transport and Climate Change: Cutting Carbon Dioxide Emissions from Cars.* Friends of the Earth, London

Hood C and Jackson M (1991) *Administrative Argument.* Dartmouth, Aldershot

Hope C, Parker J and Peake S (1992) A pilot environmental index for the UK in the 1980s. *Energy Policy* 20(4), 335–343

Houghton J T, Jenkins G J and Ephraums J J (Eds) (1990) *Climate Change: The IPCC Assessment.* Cambridge University Press, Cambridge

Huisingh D (1988) *Good Environmental Practice – Good Business Practices.* Wissenschaftszentrum fur Sozialforschung, Berlin

Inhaber H (1975) *Environmental Indices.* Wiley, London

Institute of Directors (1993) *Members Opinion Survey: Environment.* Director Publications, London

Irvine S and Ponton A (1988) *A Green Manifesto.* Optima, London

Irwin A (1994) Science's social standing. *Times Higher,* 30 September

Jackson T (1992) *Efficiency without Tears: No-Regrets Policy to Combat Climate Change.* Friends of the Earth, London

Jacobs M (1991) *The Green Economy.* Pluto, London

Jacobs M (1995) *Sustainability and Socialism.* SERA, London

Jacobs M and Stott M (1992) Sustainable development and the local economy. *Local Economy* 7(3)

Jenkins T and McLaren D P (1994) *Jobs and the Environment.* Friends of the Earth, London

Joseph S (1992) The politics of transport and the environment. *ECOS* 13(4), 2–6

Kaplan R and Talbot J (1988) Ethnicity and preference for natural settings: a review of recent findings. *Landscape and Urban Planning* 15, 107–117

Karas J H W, Roberts S and Weir F (1993) *Friends of the Earth Response to the Department of the Environment's Discussion Document on Climate Change: Our National Programme for CO$_2$ Emissions.* FOE, London

Keith M and Pile S (1993) *Place and the Politics of Identity.* Routledge, London

Kellert S R (1984) Urban American perceptions of animals and the natural environment. *Urban Ecology* 8

Kellert S R and Westerfelt M O (1983) *Children's Attitudes, Knowledge and Behaviours toward Animals.* Government Printing Office, Washington DC

Kelly P (1984) *Fighting for Hope.* Chatto & Windus, London

Keyes J, Munt I and Riera P (1993) The control of development in Spain. *Town Planning Review,* 64(1), 47–63

Kidd C V (1992) The evolution of sustainability. *Journal of Agricultural and Environmental Ethics* 5(1), 1–26

King Y (1983) Toward an ecological feminism and a feminist ecology. In J Rothschild (Ed.) *Machina Ex Dea.* Pergamon, Oxford

King Y (1989) The ecology of feminism and the feminism of ecology. In J Plant (Ed.) *Healing the Wounds.* Green Print, London

Kirby A (1981) The Politics of Location
Kirklees Borough Council (1989) *The State of the Environment in Kirklees.* Kirklees MBC, Huddersfield
Krause F, Bach W and Koomey J (1989) *Energy Policy in the Greenhouse.* International Project for Sustainable Energy Paths, California
Krauss C (1993) Blue-collar women and toxic-waste protests: the process of politicisation. In R Hofrichter (Ed.) *Toxic Struggles: The Theory and Practice of Environmental Justice.* New Society Publishers, Philadelphia
Krishnarayan V and Thomas H (1993) *Ethnic Minorities and the Planning System.* RTPI, London
Labour Party (1994) *In Trust for Tomorrow: Report of the Labour Party Policy Commission on the Environment.* Labour Party, London
Lang R (1979) Environmental information in a planning/management context. In *Proceedings of the 2nd Conference on the Applications of Classification in Canada,* pp. 285–294
Lewenhak S (1992) *The Revaluation of Women's Work.* Earthscan, London
Lex Service (1994) *1994 Lex Report on Motoring: The Consumer View.* Lex Service, London
Lindblom C and Cohen D (1979) *Usable Knowledge.* Yale University Press, New Haven
Local Government Management Board (1990) *Environmental Practice in Local Government.* LGMB, Luton
Local Government Management Board (1992) *Environmental Practice in Local Government,* 2nd edn. LGMB, Luton
Local Government Management Board (1993) *Towards Sustainability: The EC's Fifth Action Programme on the Environment: A Guide for Local Authorities.* LGMB, Luton
Local Government Management Board (1994a). *The Eco-Management and Audit Scheme: A Guide for Local Authorities.* HMSO, London
Local Government Management Board (1994b) *Local Agenda 21 Roundtable Guidance 1: Community Participation in Local Agenda 21.* LGMB, Luton
Local Government Management Board (1994c) *Local Agenda 21 Roundtable Guidance 2: North/South Linking for Sustainable Development.* LGMB, Luton
Local Government Management Board (1994d) *Local Agenda 21 Roundtable Guidance 3: Greening the Local Economy.* LGMB, Luton
Local Government Management Board (1994e) *Local Agenda 21 Roundtable Guidance 4: Educating for a Sustainable Local Authority.* LGMB, Luton
Local Government Management Board (1994f) *Local Agenda 21 Roundtable Guidance 5: Planning, Transport and Sustainability.* LGMB, Luton
Local Government Management Board (1994g) *Local Agenda 21 Roundtable Guidance 6: Green Purchasing and CCT.* LGMB, Luton
London Borough of Hackney (1994) *Action for the Environment: A Sustainable Development Plan for Hackney.* Hackney Council
Lowe P and Goyder J (1983) *Environmental Groups in Politics.* Allen and Unwin, London
Lyons E and Breakwell G M (1992) Factors predicting environmental concern and indifference in 13–16 year olds. *Environment and Behavior*
Lyotard J F (1982) *The Post-modern Condition.* Manchester University Press, Manchester
MacPherson C B (1962) *The Political Theory of Possessive Individualism.* Oxford University Press Oxford
Major J (1991) *Speech to Sunday Times Conference,* 8 July 1991, London
Majocchi A (1994) *The Employment Effects of Eco-taxes: A Review of Empirical Models and Results.* Paper presented at the OECD Workshop on Implementation of Environmental Taxes, Paris, February 1994

Majone G (1989) *Evidence, Argument and Persuasion in the Policy Process.* Yale University Press, New Haven

Malik S (1992) Colours of the countryside – a whiter shade of pale. *ECOS* 13(4), 33–40

Marlow D (1992) Eurocities: from urban networks to a European urban policy. *Ekistics* 352/353, 28

Marshall T C (1992a) A review of recent developments in European environmental planning. *Journal of Environmental Planning and Management* 35(2), 129–144

Marshall T (1992b) Environmental Sustainability: London's unitary development plans and strategic planning. School of Land Management and Urban Policy Occasional Paper OP/4/92, South Bank University

Martin D (1990) Regional/spatial planning: a task for the European Community in the 90s. In *PTRC Proceedings, 18th Annual Summer Meeting.* PTRC, London

Marvin S J (1992) Towards sustainable urban environments: the potential for Least-Cost Planning approaches. *Journal of Environmental Planning and Management* 35(2), 193–200

Matthews J A (1981) *Social Identity and Cognition of the Environment.* Unpublished PhD Thesis, University of Sheffield

Matthews J A (1983) Environmental change and community identity. In G M Breakwell (Ed.) *Threatened Identities.* Wiley, Chichester

McConnell S (1981) *Theories for Planning.* Heinemann, London

McLaren D P (1993a) Compact or dispersed: dilution is not the solution. In *Built Environment* 18(4), 268–284

McLaren D P (1993b) The future for EIA: meeting demands for sustainable development. *Paper to Environmental Assessment conference.* Aston University, Birmingham, 13 May

Meadows D H *et al.* (1972) *The Limits to Growth.* Pan, London

Meissner W (1986) Employment, income and welfare implications of environmental policy. In A Schnaiberg, N Watts and K Zimmerman (Eds) *Distributional Conflicts in Environmental-Resource Policy.* Blackmore Press, Shaftesbury, pp. 38–48

Mellor M (1992a) *Breaking the Boundaries.* Virago, London

Mellor M (1992b) EcoFeminism and ecosocialism: dilemmas of essentialism and materialism. *Capitalism, Nature, Socialism* 3(2), 1–20

Mellor M (1992c) Green politics: ecofeminist, ecofeminine or ecomasculine? *Environmental Politics* 1(2), 229–251

Mellor M (1993) Building a new vision: feminist green socialism. In R Hofrichter (Ed.) *Toxic Struggles: The Theory and Practice of Environmental Justice.* New Society Publishers, Philadelphia

Merchant C (1980) *The Death of Nature.* Harper & Row, New York

Micklewright S (1987) *Who are the New Conservationists?* Discussion papers in conservation, 46, University College, London

Mies M and Shiva V (1993) *Ecofeminism.* Zed, London

Mies M, Bennholdt-Thompson, V and von Werlhof C (1988) *Women: The Last Colony.* Zed, London

Miles R (1988) *Women's History of the World.* Penguin, London

Miles R (1993) *Racism after Race Relations.* Routledge, London

Ministry of Transport, Public Works and Water Management (1992) *Structured Scheme for Traffic and Transport. Bicycles First, the Bicycle Master Plan.* Information Department, The Hague

MINTEL (1989) *Bicycles.* Mintel International Group, September 1989

Mitchell R C (1980) *Public Opinion on Environmental Issues, Results of a National Public Opinion Poll.* CEQ, DOA, DOE and EPA, Govt Printing Office, Washington, DC

Mitter S (1986) *Common Fate, Common Bond.* Pluto, London

Mohai P (1985) Public concern and elite involvement in environmental conservation. *Social Science Quarterly.* 66 (December), 820–838

Morgan G, Fennell J and Farrer J (1993) Authorities struggling to deliver sustainable plans. *Planning* 1047, 20–21

Morphet J (1994a) *Agenda 21 and Towards Sustainability: Translating Rio into European Action.* LGMB, Luton

Morphet J (1994b) Regions arise. *Town and Country Planning* 63(1), 18–19

Moss H (1994) Consumption and fertility. In W Harcourt (Ed.) *Feminist Perspectives on Sustainable Development.* Zed, London

Myerson G and Rydin Y (1994) 'Environment' and planning: a tale of the mundane and the sublime. *Society and Space* 12, 437–452

Naess A (1973) The shallow and the deep, long-range ecology movement. *Inquiry* 16, 95–100

Næss P (1993) Can urban development be made environmentally sound? *Journal of Environmental Planning and Management* 36(3), 309–333

Nash C and Hopkinson P (1992) Transport appraisal – the loaded dice. *ECOS* 13(4), 6–10

NCC (Nature Conservancy Council) (1989) *Guidelines for the Selection of Biological SSSIs.* NCC, Peterborough

National Audit Office (1993) *Pergau Hydro-electric Project.* HMSO, London

Newman P (1994) Killing legally with toxic waste: women and the environment in the United States. In V Shiva (Ed.) *Close to Home: Women Reconnect Ecology, Health and Development Worldwide.* New Society Publishers, Philadelphia

Nicolaides P (Ed.) (1993) *Industrial Policy in the European Community: A Necessary Response to Economic Integration.* Martinus Nijhoff, Dordrecht

Nordgaard R (1988) Sustainable development: a co-evolutionary view. *Futures* 20(6), 24–42

O'Brien M (1993) Environmental Culture? The social organisation and disorganisation of the environment. *Paper presented to the Interdisciplinary Research Network on Environment and Society Conference*, Sheffield

ODA (1993) Press Release, 7 December 1993

OECD (Organisation for Economic Cooperation and Development) (1985) *Environmental Policy and Technical Change.* OECD, Paris

OECD (1987) *The Promotion and Diffusion of Clean Technologies in Industry.* OECD Environment Monograph 9, OECD, Paris

OECD (1991) *Environmental Indicators: A Preliminary Set.* OECD, Paris

OFFER (Office of Electricity Regulation) (1993) *Electricity Distribution: Price Control, Reliability and Customer Service.* Consultation Paper, Birmingham, OFFER

O'Riordan T and Cameron J (Eds) (1994) *Interpreting the Precautionary Principle.* Cameron May, London

O'Riordan T and Turner R K (Eds) (1983) *An Annotated Reader in Environmental Planning and Management.* Pergamon Press, Oxford

Owens S (1994) Land, limits and sustainability: a conceptual framework and some dilemmas for the planning system. *Transactions of the Institute of British Geographers*, 19(4), 439–456

Peach C (1986) Patterns of Afro-Caribbean migration and settlement in Great Britain: 1945–1981. In C Brock (Ed.) *The Caribbean in Europe.* Cass, London

Pearce B J (1992) The effectiveness of the British land-use planning system. *Town Planning Review* 63(1), 13–28

Pearce D (1993) *Blueprint 3: Measuring Sustainable Development.* Earthscan, London

Pearce D and Brisson I (1993) BATNEEC: the economics of technology based standards. *Oxford Review of Economic Policy* 9(4), 24–40

Pearce D, Markandya A and Barbier E (1989) *Blueprint for a Green Economy.* Earthscan, London

Pepper D (1993) *Eco-socialism: From Deep Ecology to Social Justice.* Routledge, London

Pietila H (1987) Alternative development with women in the North. *Paper presented at the Third International Interdisciplinary Congress of Women,* Dublin, 6–10 July. Also published in J Galtung and M Friberg (Eds) *Alternative Akademilitteratur,* Stockholm, 1986

Plant J (Ed.) (1989) *Healing the Wounds.* Green Print, London

Plumwood V (1993) *Feminism and the Mastery of Nature.* Routledge, London

Port G N J (1980) Integrated global monitoring of the environment. In *Proceedings of the Symposium on the Development of Multimedia Monitoring of Environmental Pollution.* World Meteorological Organisation, Geneva, pp. 183–195

Porter G and Welsh Brown J (1991) *Global Environmental Politics.* Westview, Oxford

Poulton M C (1991) The case for a positive theory of planning. Part 1: What is wrong with planning theory? *Environment and Planning B: Planning and Design* 18(2), 223–232

Rao B (1989) Struggling for production conditions and producing conditions of emancipation: women and water in rural Maharashtra. *Capitalism, Nature, Socialism* 2 (Summer)

Rau B (1991) *From Feast to Famine.* Zed, London

Readc E (1987) *British Town and Country Planning.* Open University Press, Milton Keynes

Rice T (Ed.) (1993) *The Rainforest Harvest.* Friends of the Earth, London

Rich B (1994) *Mortgaging the Earth: The World Bank, Environmental Impoverishment and the Crisis of Development.* Earthscan, London

Rijksplanologische Dienst (1991) *Perspectives in Europe: A Survey of Options for a European Spatial Policy.* RPD, The Hague

Riley D (1988) *Am I that Name? Feminism and the Category of 'Women' in History.* Macmillan, London

Roberts S (1992) *Energy for a Future.* Friends of the Earth, London

Rovinsky F Y (1982) Second international symposium on integrated global monitoring of environmental pollution. *Environmental Monitoring and Assessment* 2(4), 361–367

Rowell T (1991) *SSSIs: A Health Check.* Wildlife Link, London

Rowntree, Joseph Foundation (1994) Joseph Rowntree Foundation Findings, *Housing Research,* 110

Royal Commission on Environmental Pollution (1992) *Freshwater Quality,* HMSO, London

Rydin Y and Myerson G (1989) Explaining and interpreting ideological effects: a rhetorical approach to green belts. *Society and Space* 7, 463–479

Scharer B (1993) Technologies to clean up power plants. *Staub-Rheinhaltung der Luft,* 53, 87–92

Schmidheiny S (1992) *Changing Course: A Global Business Perspective on Development and the Environment.* MIT Press, Cambridge, MA

Schon D A (1983) *The Reflective Practitioner.* Basic Books, New York

Schon D A (1987) *Educating the Reflective Practitioner.* Jossey-Bass, San Francisco

Schreiner O (1978) *Woman and Labour.* Virago, London. First published in 1911

Scottish Natural Heritage (1993) *Annual Report.* Scottish Natural Heritage, Edinburgh.

Seager J (1993) *Earth Follies: Feminism, Politics and the Environment.* Earthscan, London

Selman P (1992) *Environmental Planning.* Paul Chapman, London

Sen G (1994) Women, poverty and population: issues for the concerned environmentalist. In W Harcourt (Ed.) *Feminist Perspectives on Sustainable Development.* Zed, London

Sen G and Grown C (1987) Development, crises and alternative visions. *Monthly Review,* New York

Shiva V (1989) *Staying Alive.* Zed Press, London

Shiva V (1993a) *Monocultures of the Mind.* Zed Books, London

Shiva V (1993b) Recovering the real meaning of sustainability. In D E Cooper and J A Palmer (Eds) *The Environment in Question: Ethics and Global Issues.* Routledge, London, pp. 187–193

Shiva V (Ed.) (1994) *Close to Home: Women Reconnect Ecology, Health and Development Worldwide.* New Society Publishers, Philadelphia

Simmie J (1993) *Planning at the Crossroads.* UCL Press, London

Sivanandan A (1988) The new racism. *New Statesman and Society* 1(22), 8–9

Skellington R (1992) *Race in Britain Today.* Sage, London

Smith S (1989) *The Politics of Race and Residence.* Polity, Cambridge

Spackman M (1992) Trends in traffic and emissions. *Paper presented to the 1992 Cambridge Econometrics Conference, Transport, Communications and the Economy: Imagining the 21st Century*

Steer A and Lutz E (1993) Measuring environmentally sustainable development. *Finance and Development* December 1993, pp. 20–23

Stevenson R (1993) Thinking, believing and persuading: some issues for environmental activists. In D E Cooper and J A Palmer (Eds) *The Environment in Question: Ethics and Global Issues.* Routledge, London, pp. 212–223

Steward F (1991) Citizens of planet earth. In G Andrews (Ed.) *Citizenship.* Lawrence and Wishart, London

Stokes G, Pickup L, Meadowcroft, S, Kenny F and Goodwin P (1990) *Bus Deregulation: The Metropolitan Experience.* Association of Metropolitan Authorities, London

SWCC (1990) *Second World Climate Conference.* World Meteorological Organisation, Geneva

Swyngedouw E (1992) The Mammon Quest. Glocalisation, interspatial competition and the monetary order: the construction of new scales. In M Dunford and G Kafkalas (Eds) *Cities and Regions in the New Europe: The Global–Local Interplay and Spatial Development Strategies.* Belhaven, London

Tajfel H (1982) *Social Identity and Intergroup Relations.* Cambridge University Press, Cambridge

Taylor D (1989) Blacks and the environment: towards an explanation of the concern and action gap between blacks and whites. *Environment and Behaviour* 21(2), 175–205

Taylor D (1993) Minority environmental activism in Britain: from Brixton to the Lake District. *Qualitative Sociology* 16(3), 263–295

Tolba M, El-Kholy O A, El-Hinnawi E, Holdgate M W, McMichael D F and Munn R E (Eds) (1992) *The World Environment 1972–1992.* UNEP, Chapman and Hall, London

Treweek J (1993) Ecological assessment of proposed road developments: a review of Environmental Statements. *Journal of Planning and Environmental Management* 36(3), 295–307

Tuxworth B (1994) Local authorities blaze the Agenda 21 trail. *Town and Country Planning* August 1994

Tuxworth B and Carpenter C (1995) *Local Agenda 21*. Surrey 1994/5, Local Government Management Board, Luton

UNEP (1991) *Benchmark Corporate Environmental Survey*. United Nations, New York

UNEP–UK (1992) *Good Earth Keeping*, UNEP–UK, London

United Nation Conference on Environment and Development (1992) *Agenda 21*. UNCED, Geneva

Unwin N (1993) *The Determinants of Cycling Behaviour*. Transport and Health Study Group

Urquhart C (1994) Gas bill to pay. *The Guardian* 30 November

USEPA (United States Environmental Protection Agency) (1988) *Toxics in the Community*. USEPA, Washington, DC

Van Ardsol M D Jr, Sabagh G and Alexander F (1965) Reality and the perception of environmental hazards. *Journal of Health and Human Behaviour* 5

Walley W J and Judd S (Eds) (1993) River water quality management and control. *Proceedings of the Freshwater Europe Symposium*. Aston University, Birmingham

Ward D (1989) *Poverty, Ethnicity and the American City 1840–1925: Changing Conceptions of the Slum and the Ghetto*. Cambridge University Press, Cambridge

Waring M (1989) *If Women Counted*. Macmillan, London

Warren Spring Laboratory (1993) *Report of the Current Work of the Warren Spring Laboratory*. Investigation of Air Pollution Standing Conference

Weizsacker E U von (1994) *Earth Politics*. Zed, London

Weizsacker E U von and Jesinghaus J (1992) *Ecological Tax Reform*. Zed, London

Whatmore S and Boucher S (1993) Bargaining with nature: the discourse and practice of environmental planning gain. *Transactions of the Institute of British Geographers* 18(2), 166–178

Whittaker M (1986) Cholinesterase. *Monographs in Human Genetics*, Vol. II. Karger, Basel

Williams R (1988) The European Communities directive on environmental impact assessment. In M Clark and J Herington (Eds) *The Role of EIA in the Planning Process*. Mansell, London

Williams R (1990) European spatial planning strategies and environmental planning. In G Ashworth and P Kivell (Eds) *Land, Water and Sky: European Environmental Planning*, Geo Pers, Groningen, pp. 9–17

Williams R (1991) Placing Britain in Europe: four issues in spatial planning. *Town Planning Review* 62(3), 331–340

Williams R (1994) Spatial planning for an integrated Europe. In J Lodge (ed.) *The European Community and the Challenge of the Future*. Pinter, London

Willson J S and Greeno J L (1993) Business and the environment: the shape of things to come. *Prism*, Third Quarter, 5–18

Wood B and Williams R (Eds) (1992) *Industrial Property Markets in Western Europe*. Spon, London

Wood C and Jones C (1991) *Monitoring Environmental Assessment and Planning*. HMSO, London

World Commission on Environment and Development (1987) *Our Common Future*. Oxford University Press

Yanarella E J and Levine R S (1992a) The Sustainable Cities Manifesto: pretext, text and post-text. *Built Environment* 18(4), 301–313

Yanarella E J and Levine R S (1992b) Does sustainable development lead to sustainability? *Futures* 24(8), 759–774

Yiftachel O and Hedgcock D (1993) Urban social sustainability: the planning of an Australian city. *Cities* 10(2), 139–157

Index